中等职业教育改革发展示范学校创新教材

路由器交换机配置教程（项目式）

Router and Switch
Configuration Tutorial

李士山 ◎ 主编
张秀芹 张学义 杨光 ◎ 副主编

人民邮电出版社

北京

图书在版编目（CIP）数据

路由器交换机配置教程：项目式 / 李士山主编. --北京：人民邮电出版社，2014.7（2023.12重印）
中等职业教育改革发展示范学校创新教材
ISBN 978-7-115-35040-4

Ⅰ. ①路… Ⅱ. ①李… Ⅲ. ①计算机网络－路由选择－中等专业学校－教材②计算机网络－信息交换机－中等专业学校－教材 Ⅳ. ①TN915.05

中国版本图书馆CIP数据核字(2014)第073143号

内 容 提 要

本书以培养学生的路由器交换机配置技能为核心，以工作过程为导向，以思科 PACKET TRACER 模拟软件为支持，在一定程度上缓解了实训设备不足的情况。书中详细介绍了路由器交换机的配置命令及其过程，能够满足中等职业学校计算机网络技术路由交换配置的教学需求。

本书以工作过程为导向，采用项目教学的方式组织内容。全书包括 7 个项目，前 6 个项目由简单到复杂，每个项目均由实训目的、实训任务、预备知识、实训步骤等部分组成；第 7 个项目精心选择四套竞赛题，借以开阔视野。通过学习和实训，学生不仅能够掌握路由交换设备配置的基础知识，而且能够熟练掌握路由器交换机的配置技能，满足一般中小型企业对网络技术人员的需求。

本书可作为中等职业技术学校计算机网络技术专业的教学实训用书，也可作为参加省、市、全国企业网搭建项目技能大赛学生训练的参考教材，并可供网络管理人员参考、学习之用。

◆ 主　编　李士山
　　副 主 编　张秀芹　张学义　杨　光
　　责任编辑　桑　珊
　　责任印制　焦志炜

◆ 人民邮电出版社出版发行　北京市丰台区成寿寺路11号
　　邮编 100164　电子邮件 315@ptpress.com.cn
　　网址 http://www.ptpress.com.cn
　　固安县铭成印刷有限公司印刷

◆ 开本：787×1092　1/16
　　印张：16.5　　　　　　　　　　2014年7月第1版
　　字数：422千字　　　　　　　　　2023年12月河北第19次印刷

定价：39.80元

读者服务热线：(010)81055256　印装质量热线：(010)81055316
反盗版热线：(010)81055315

青岛开发区职业中专示范校建设系列教材编委会

主　任： 崔秀光

副主任： 侯方奎　薛光来　李本国　杨逢春　姜秀文　王济彬

委　员： 赵贵森　张学义　韩维启　丁奉亮　邹　蓉　王志周

　　　　　 张元伟　王　部　张　栋　薛正香　王本强　李玉宁

　　　　　 赵　萍　彭琳琳　李士山　荆建军　殷茂胜　宋　芳

　　　　　 徐锡芬　毛　慧　王景涛　郭晓宁　刘　萍　王云红

　　　　　 何　彬　杜召强　潘进福　朱秀萍　焦风彩　赵　丽

　　　　　 于雅婷　王莉莉

校外参编人员（按姓氏笔画排序）

丁海萍　　上汽通用五菱汽车股份有限公司青岛分公司　经理

王　涛　　青岛华瑞汽车零部件有限公司车间　主任

孙义振　　青岛澳柯玛洗衣机有限公司　副总经理、高级工程师

孙红菊　　北京络捷斯特科技发展有限公司　副总经理

孙　斌　　青岛汇众科技有限公司　主任

张正明　青岛来易特机电科技有限公司　经理

吴向阳　山东工艺美术学院　教授/系主任

邵昌庆　青岛金晶玻璃有限公司　副总经理、高级工程师

秦　朴　青岛城市名人酒店　总经理

徐增佳　上汽实业有限公司（青岛分公司）　总经理助理

薛培财　青岛旭东工贸有限公司　经理

前　言

近年来，职业教育转变了以课堂教学为中心的传统职业教育模式，大力推行工学结合、校企合作、顶岗实习的人才培养新模式，注重学生职业技能素质方面的培养，职业教育办学质量逐步得到提高。

本书贴近中等职业学校教学实际，以工作过程为导向，以思科 PACKET TRACER 模拟软件为支持，采用项目教学的方式组织内容。主要内容包括七个由简单到复杂的路由交换设备配置项目。

项目 1 网络基础

本项目介绍了常见的网络设备及其功能作用；讲解双绞线的制作技艺。实训连接各种常见的网络设备，组建简单的计算机网络。

项目 2 交换机的基本配置

本项目详细讲解交换机的基本配置命令；划分 VLAN 实现部门间的隔离；生成树的原理和基本配置；配置交换机端口安全。

项目 3 配置路由器

本项目讲解路由器的原理；实训配置单臂路由，实现各个 VLAN 间互相通信；了解静态路由和默认路由的应用场合；实训配置静态路由、默认路由和浮动静态路由。

项目 4 动态路由协议

本项目介绍动态路由和静态路由的特点、区别，实训动态路由协议 RIP 和 OSPF 的配置。

项目 5 构建安全的网络

本项目讲解 ACL 的语法规则和执行过程，实训标准 ACL、扩展 ACL 的配置方法。学习并配置 PPP　PAP 认证和 PPP　CHAP 认证。

项目 6 连入 Internet

本项目实训静态 NAT、动态 NAT 和 NAPT 的配置，通过 IPSEC VPN 构建安全的互联环境，让我们共享网上冲浪的乐趣！

项目 7 竞赛试题

本项目精心选择 4 套竞赛试题。增强综合运用所学知识的能力，使学生对各级技能大赛有初步认识。

前 6 个项目由实训目的、实训任务、预备知识、实训拓扑、实训步骤等部分组成。实训目的明确提出实训达成的目标；实训任务列举出学生实训应完成哪些任务；预备知识部分讲解实训所需的基本知识，为学生奠定理论基础；实训拓扑旨在训练读图能力，使学生能正确连接设备搭建网络；在实训步骤部分，按照实训实际进行过程，全面给出配置命令，并介绍配置命令的功能作用，使学生动手实践时能够明确每条配置命令的作用，不仅要知其然，更要知其所以然。实训过程中巧妙设置了要学生自己验证测试的环节，让学生想一想为什么，加深学生对知识的理解。第七个项目精心选择 4 套竞赛试题，开阔视野。

本书的参考学时为 60~70 学时，建议采用理论实践一体化的教学模式。

由于编者水平和经验有限，书中难免有欠妥和不足之处，恳请读者批评指正。

编 者

2013 年 10 月

目　录

项目 1　网络基础 1

　　任务 1.1　熟悉 Packet Tracer 2
　　任务 1.2　认识网络设备 8
　　任务 1.3　制作双绞线 10
　　任务 1.4　解析 OSI 参考模型和 TCP/IP 15

项目 2　交换机的基本配置 34

　　任务 2.1　交换机配置方式及基本操作 35
　　任务 2.2　设置交换机密码 40
　　任务 2.3　划分 VLAN，实现部门间隔离 45
　　任务 2.4　跨交换机实现同一 VLAN 的主机通信 50
　　任务 2.5　交换机端口安全 54
　　任务 2.6　配置快速生成树 57

项目 3　配置路由器 64

　　任务 3.1　简单配置路由器 65
　　任务 3.2　配置路由器密码 69
　　任务 3.3　单臂路由 76
　　任务 3.4　静态路由和默认路由 79
　　任务 3.5　浮动静态路由 87

项目 4　动态路由协议 93

　　任务 4.1　动态路由协议 RIPV2 94
　　任务 4.2　单区域的 OSPF 100
　　任务 4.3　OSPF 基于区域的 MD5 认证 108
　　任务 4.4　多区域 OSPF 114

| 任务 4.5 | 路由重发布 | 128 |
| 任务 4.6 | 路由选择原则 | 134 |

项目 5　构建安全的网络 144

任务 5.1	PPP PAP 认证	145
任务 5.2	配置 PPP CHAP 认证	149
任务 5.3	标准 ACL	153
任务 5.4	扩展 ACL	160

项目 6　连入 Internet 169

任务 6.1	静态 NAT	170
任务 6.2	动态 NAT 和 NAPT	175
任务 6.3	IPSECVPN	182
任务 6.4	配置 IPSEC VPN 和 NAT	188
任务 6.5	配置 GRE VPN	196
任务 6.6	配置 IPSEC OVER GRE	203

项目 7　竞赛试题 211

2010 年青岛市中等职业学校企业网搭建与应用技能比赛 212
2011 年全国中职技能大赛企业网搭建及应用模拟试题 221
2012 年青岛市中职技能大赛企业网搭建及应用模拟试题 232
2013 年青岛市中职技能大赛企业网搭建及应用模拟试题 243

项目 1

网络基础

　　计算机网络就是把分散的具有独立功能的计算机通过连接介质互连起来，按照网络协议进行数据通信，实现资源共享的一种组织形式。构建一个简单的计算机网络，所需要的硬件通常有计算机、路由器、交换机、传输介质（双绞线、光纤、同轴电缆等）、网卡、连接器件等。

　　通过本项目的实训，我们学习 Cisco 公司发布的模拟软件 Packet Tracer 的使用；认识常见的网络设备，了解其功能作用；掌握双绞线的制作技艺，学会连接各种常见的网络设备和组建简单的计算机网络；解析 OSI 模型的原理，领悟数据传送过程中如何封包拆包。

任务 1.1　熟悉 Packet Tracer

【实训目的】

学习 Packet Tracer 的常用操作，为后续实训打好基础。

【实训任务】

1. 熟悉 Packet Tracer 界面，了解各组成部分的功能作用。
2. 添加路由器交换机等设备，并用线缆正确连接，构建实训拓扑。
3. 练习在命令行模式下配置路由器交换机。
4. 在模拟方式下查看 PDU 信息。
5. 汉化 Packet Tracer。

【实训设备】

Packet Tracer5.3。

【实训步骤】

步骤 1　认识 Packet Tracer 界面。

Packet Tracer 是由 Cisco 公司发布的一个辅助学习工具，主要作用是为学习思科网络课程的初学者去设计、配置、排除网络故障提供网络模拟环境。用户可以在软件的图形用户界面上直接通过拖曳方法建立网络拓扑，并可提供数据包在网络中行进的详细处理过程，观察网络实时运行情况。

运行 Packet Tracer，进入其工作界面。思科 Packet Tracer 工作界面包括菜单、工具栏、工作区、设备区等几个部分，如图 1-1-1 所示。

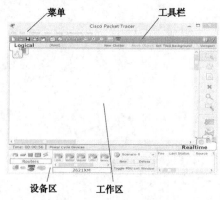

图 1-1-1　Packet Tracer 工作界面

步骤 2　添加网络设备。

在设备类型区列举着许多种类的硬件设备，图 1-1-2 中从左至右，从上到下依次为路由器、交换机、集线器、无线设备、设备之间的连线（Connections）、终端设备、仿真广域网和自定义设备（Custom Made Devices）。

图 1-1-2　Packet Tracer 设备区

当鼠标指向设备类型区某设备时，其下方会显示出设备名称。当单击某一类型设备时，在设备型号区会出现相应类型设备的型号列表。这时单击需要的设备型号，然后在工作区单击，即添加了该型号的网络设备。重复以上过程，将我们所需要的设备一一添加到工作区。

步骤 3　设备连接。

在图 1-1-2 所示的 Packet Tracer 设备区中，单击线缆按钮，再在图 1-1-3 所示的线缆选择区中单击某一线型，然后在工作区需要连接的设备上单击鼠标左键，选择要连接的接口，再移动鼠标到要连接的目标设备上单击左键，选择要连接的接口。这样，两个设备就连接好了。

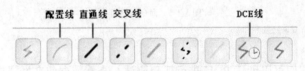

图 1-1-3　Packet Tracer 的线型

我们常用到配置线、直通线、交叉线和 DCE 线这四种线型。配置线一端连接电脑的串口，另一端连接设备的 console；直通线用于连接不同类型的设备，比如路由器和交换机、计算机和交换机等；交叉线用于连接相同类型的设备，比如路由器与路由器以太网口的连接、交换机与交换机的连接等；DCE 线用于路由器串口之间的连接。这里需要注意的是：

①计算机和路由器以太口间的连接用交叉线；

②用 DCE 线首先连接的路由器为 DCE，相连的路由器串口需要配置时钟。

步骤 4　更换设备或线缆。

当需要更换设备时，我们单击工作区右上角区域中的删除按钮。删除掉设备，再新增设备即可。同样的道理，我们也可以更换线缆类型。

这样，我们在工作区就建立起了一个完整的网络拓扑图，如图 1-1-4 所示。

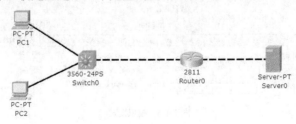

图 1-1-4　网络拓扑结构图

步骤 5　配置设备。

单击要配置的设备,在弹出的对话框中单击 CLI 选项卡,即进入该设备的命令行界面,如图 1-1-5 所示。

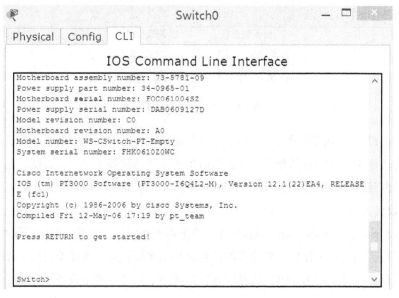

图 1-1-5　命令行界面

步骤 6　模拟模式。

在模拟模式下,我们可以创建 PDU 进行抓包测试,查看数据包在网络中的传输情况。

(1)单击 Packet tracer 右下角的 Simulation 按钮,进入模拟模式,如图 1-1-6 所示。

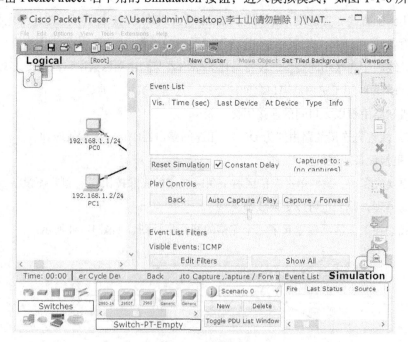

图 1-1-6　模拟模式

（2）单击 Add Complex PDU (c)按钮,再单击发出数据包的源设备,弹出 Create Complex PDU 对话框,如图 1-1-7 所示。

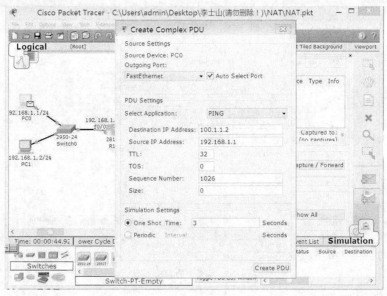

图 1-1-7 创建 PDU

（3）选择输出接口和要运行的指令,输入目标 IP 地址和源 IP 地址、序列号,时间等参数后,单击 Create PDU 按钮。

（4）单击 Auto Capture/Play 按钮,可以看到信封状的数据包从源设备到目的设备之间的传输过程,如图 1-1-8 所示。

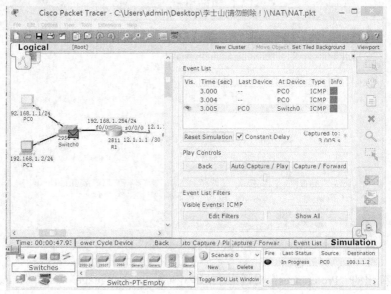

图 1-1-8 数据包传输过程

（5）单击 Event list 中 Info 列下的某一彩色方块，可以查看设备上的 PDU 信息。查看的方法将在后续章节中进行详细讲解，如图 1-1-9 所示。

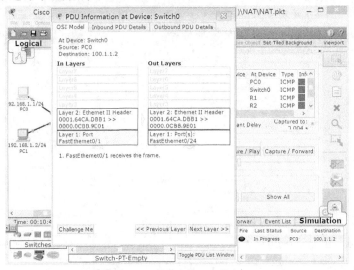

图 1-1-9　设备上的 PDU 信息

步骤 7　汉化 Packe Tracer。

（1）针对英语基础薄弱的用户，Packe Tracer 提供了汉化包。我们只要将下载的汉化文件 chinese.ptl 存放到软件的 languages 文件夹中，然后单击"Options"菜单中的"Preferences……"即可，如图 1-1-10 所示。

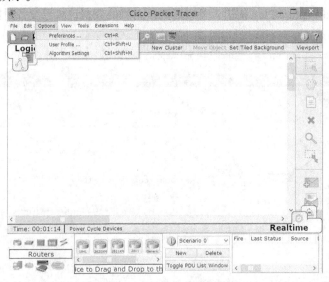

图 1-1-10　配置 Packe Tracer

（2）在弹出的"Preferences"对话框中，单击"Select Language"下的"chinese.ptl"，再单击"Change Language"按钮，如图 1-1-11 所示。

图 1-1-11　设置 Packe Tracer 语言

（3）弹出的对话框中提示下次启动软件时更改语言，单击"OK"按钮即可，如图 1-1-12 所示。

图 1-1-12　确认汉化 Packe Tracer

（4）汉化的 Packe Tracer。

重启 Packe Tracer，就会看到汉化后的 Packe Tracer 了，如图 1-1-13 所示。

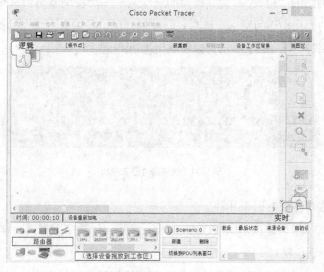

图 1-1-13　汉化的 Packe Tracer 界面

Packe Tracer 功能强大，是我们学习计算机网络技术的有力帮手。随着后续实训的进行，读者一定可以越来越熟练地使用它，并深刻体会到它的好处。

任务 1.2　认识网络设备

【实训目的】

认识常用的网络设备和连接线缆。

【实训任务】

1．实际观察交换机、路由器等设备的外观，识别这些设备的网络连接接口。
2．识别用于连接设备的线缆。
3．观察一个实际的网络，认识其中的网络设备及其连接线缆和连接方式。

【实训设备】

锐捷 S3760E-24 交换机、集线器、锐捷 R20-18 路由器、PC 机、5 类 UTP（直通线、交叉线、反转线）若干、DTE/DCE 电缆。

【实训步骤】

步骤 1　认识路由器。

路由器是连接网络的"桥梁"。世界最大的互联网——因特网就是由成千上万的路由器将不同国家、不同地区的大大小小的网络连接在一起而形成的。我国自主生产的路由器品牌主要有华为、锐捷、神州数码等。图 1-2-1 所示为锐捷路由器 RSR20。

图 1-2-1　锐捷路由器 RSR20

路由器常见的接口主要有以下几种。

（1）高速同步串口：可连接 DDN (Digital Data Network, 数字数据网)、帧中继(Frame Relay)、X-25、PSTN(模拟电话线路)。

（2）Console 接口：该端口为异步端口，主要连接终端或运行终端仿真程序的计算机，在本地配置路由器。不支持硬件流控制。

（3）Ethernetwork 接口：用来连接以太网。有的路由器还有 Fast Ethernetwork(快速以太网)端口。

步骤 2　认识交换机。

交换机是构建局域网必不可少的网络设备。它具有自动寻址、交换处理的功能。交换机上最多的端口是自适应的 Ethernetwork 接口，可用来连接有 10/100bps 以太网接口的设备。图 1-2-2 所示为锐捷交换机 RG-S3760E。

图 1-2-2　锐捷交换机 RG-S3760E

步骤 3　认识直通线、交叉线。

双绞线是目前局域网中应用最多的连接介质，它具有性价比高和连接简便等优点。图 1-2-3 所示为制作好的双绞线。

图 1-2-3　双绞线

双绞线按照两端线序的相同或不同，分为直通线和交叉线。

直通双绞线的线序两端一般遵循 EIA/TIA 568B 标准：橙白、橙、绿白、蓝、蓝白、绿、棕白、棕。直通线用于连接不同类型的设备，如交换机和计算机、交换机和路由器等。

交叉线一端遵循 EIA/TIA 568B 标准：橙白、橙、绿白、蓝、蓝白、绿、棕白、棕。另一端遵循 EIA/TIA568A 标准：绿白、绿、橙白、蓝、蓝白、橙、棕白、棕。交叉线用于连接相同类型的设备，如交换机之间的连接，计算机和路由器的连接（路由器连接双绞线的接口和计算机一样，也是以太网卡接口），路由器之间的连接等。

步骤 4　认识 DTE/DCE 连接线缆。

DTE/DCE 连接线缆用于连接路由器的高速同步串口。图 1-2-4 所示为 DTE/DCE 连接线缆。

图 1-2-4　DTE/DCE 连接线缆

DTE/DCE 连接线缆两端一般标记有 DTE 或 DCE。其中，连接 DCE 端的接口需要配置时钟。

步骤5 认识反转线和RJ-45到DB-9适配器。

反转线缆也称控制线，线缆两端的线序相反。如果一端线序为橙白、橙、绿白、绿、蓝白、蓝、棕白、棕；另一端线序则为棕、棕白、蓝、蓝白、绿、绿白、橙、橙白。反转线一端通过RJ-45到DB-9适配器连接计算机的串行口，另一端连接网络设备的console口。反转线用于对设备的带外管理。带外管理不占用网络设备的带宽。如图1-2-5、图1-2-6和图1-2-7所示。

图1-2-5　RJ-45到DB-9适配器（1）　　图1-2-6　RJ-45到DB-9适配器（2）　　图1-2-7　反转线

步骤6 观察一个实际的设备连接图。

将图1-2-8和实训室实际网络设备连接对照，认识设备、接口、线缆，练习将路由器、交换机和计算机用相应的线缆互连起来，组成一个实训网络。

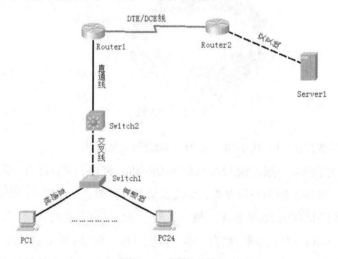

图1-2-8　设备连接图

任务1.3　制作双绞线

【实训目的】

通过制作RJ45双绞线，了解T568A/T568B标准，掌握直通性和交叉线的制作技能。

【实训任务】

1. 了解双绞线的特性及屏蔽与非屏蔽双绞线的差别。

2. 了解 T568A/T568B 标准，掌握网线的线序。
3. 制作直通线和交叉线。
4. 学会测线器的使用方法。

【预备知识】

一、双绞线概述

双绞线，也就是我们经常说的网线，是目前局域网中应用最多的传输介质。它安装简单、价格低廉，传输距离一般不超过 100 米。双绞线外有一层绝缘层，包裹着内部的八根线，每根线外部也都有绝缘层，上面标记着不同的颜色。按照橙、绿、蓝、棕分为四对，每对线相互绞在一起，使电磁辐射和外部电磁干扰降低到最小，所以叫双绞线。

双绞线可以分为屏蔽双绞线(Shielded Twisted Pair，STP)与非屏蔽双绞线(Unshielded Twisted Pair，UTP)，所谓屏蔽双绞线就是指在双绞线与外层绝缘封套之间有一个金属屏蔽层。屏蔽层可减少辐射，防止信息被窃听，也可阻止外部电磁干扰的进入，使屏蔽双绞线比同类的非屏蔽双绞线具有更高的传输速率。

二、双绞线线序标准

双绞线的制作一般遵循 EIA/TIA 标准，即 EIA/TIA 568A 和 EIA/TIA 568B。目前通用的是 EIA/TIA568B 标准。

EIA/TIA 568B 线序：

1　2　3　4　5　6　7　8
橙白、橙、绿白、蓝、蓝白、绿、棕白、棕；

EIA/TIA 568A 线序：

1　2　3　4　5　6　7　8
绿白、绿、橙白、蓝、蓝白、橙、棕白、棕。

说明："橙白"是指白线上有橙色的色点或色条的线缆，绿白、棕白、蓝白也采用同样的表示方法。

三、直通线与交叉线

直通双绞线的线序两端一般遵循 EIA/TIA568B 标准。直通线连接异型设备，如交换机和计算机、交换机和路由器等。

交叉线一端遵循 EIA/TIA568B 标准，另一端遵循 EIA/TIA568A 标准。交叉线连接同型设备，如交换机之间的连接，路由器之间的连接等。需要注意的是：计算机和路由器的连接使用的也是交叉线。

目前的网络设备大多支持端口自动翻转，不再刻意区分直通线或交叉线，连接好即可正常通信。但在我们学习的 Packet Tracer 中，交叉线和直通线是不能混淆的。

四、线序助记词

1. EIA/TIA568B 线序助记词：
橙绿蓝棕四对线，
依次拆开杂色前。
绿线蓝线要互换，
插线面朝金属面。

2. EIA/TIA568A 线序助记词：
绿橙蓝棕四对线，
依次拆开杂色前。
橙线蓝线要互换，
插线面朝金属面。

【实训拓扑】

网络拓扑结构图如图 1-3-1 所示。

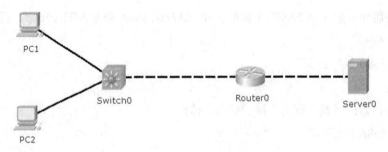

图 1-3-1 网络拓扑结构图

【实训设备】

双绞线、水晶头、网线钳和测线仪。

【实训步骤】

步骤 1 剥线。

用网线钳把双绞线的一端剪齐，然后将剪齐的一端插入到网线钳用于剥线的缺口中。顶住网线钳前挡板以后，适度握紧网线钳慢慢旋转一圈，让刀口划开双绞线的保护胶皮并剥除外皮，如图 1-3-2 所示。

图 1-3-2 剥线

步骤 2 理线。

将 4 对相互缠绕的芯线一一拆开，按照 EIA/TIA568B 标准（橙白、橙、绿白、蓝、蓝白、绿、棕白、棕）的颜色一字排列。由于线缆之前是相互缠绕着的，因此线缆会有一定的弯曲，我们使用双手的大拇指和食指捏住线缆，然后左右摇晃，将线缆尽量拉直并保持线缆平扁。再用网线钳将线的顶端剪齐，如图 1-3-3 所示。

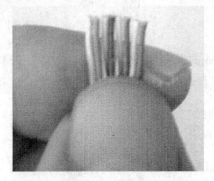

图 1-3-3 理线

步骤 3 插线。

左手捏住排列整齐的线缆，右手捏住水晶头，将水晶头的弹簧卡朝下，带金手指的一面朝向自己，再将正确排列的双绞线插入水晶头中。此处甚为关键，务必将各条芯线都插到底部，如图 1-3-4 所示。

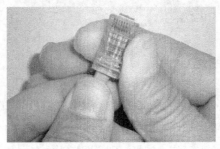

图 1-3-4 插线

步骤4 压线。

将插入双绞线的水晶头插入网线钳的压线插槽中，用力压下网线钳的手柄，如图1-3-5所示。

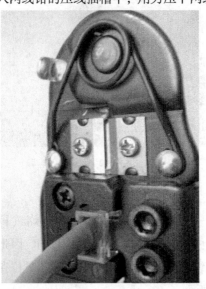

图1-3-5 压线

步骤5 制作另一端。

重复以上步骤，制作双绞线的另一端。直通线仍按EIA/TIA568B标准，如果制作交叉线，则另一端的线序按EIA/TIA568A标准（绿白、绿、橙白、蓝、蓝白、橙、棕白、棕）排列。图1-3-6所示为制作完成的双绞线。

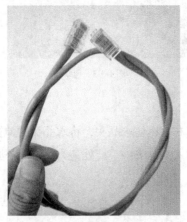

图1-3-6 制作完成的双绞线

步骤6 测试。

双绞线制作完成后，需要使用网线测试仪进行测试。将直通线的两端分别插入网线测试仪的RJ-45接口，并接通测试仪电源，如图1-3-7所示。如果测试仪上的8个绿色指示灯都顺序闪过，则说明制作成功。如果其中某个指示灯未闪烁，则说明插头中存在断路或者接触不良的现象。此时应再次对网线两端的RJ-45插头用力压一次并重新测试，如果依然不能通过测试，只能重新制作。

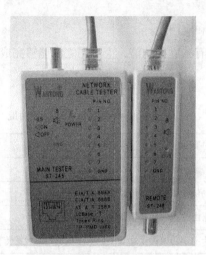

图 1-3-7 测线

测试交叉线时需要注意，测线仪灯亮的顺序和直通线不同。主控端灯亮的顺序是 1、2、3、4、5、6、7、8，测试端灯亮的顺序是 3、6、1、4、5、2、7、8。

步骤 7 连接设备。

用制作好的直通线按拓扑图所示连接计算机和交换机，用交叉线连接路由器和路由器、计算机和路由器。在教师指导下配置好路由器接口和计算机网卡 IP 地址，观察网卡指示灯、交换机接口指示灯和路由器接口指示灯的状态。

任务 1.4 解析 OSI 参考模型和 TCP/IP

【实训目的】

理解 OSI 参考模型和 TCP/IP 的原理，熟悉 Packet Tracer 的操作。

【实训任务】

1. 在 Packet Tracer 模拟模式下抓包，分析拆封包过程和每层的数据结构。
2. 理解 OSI 参考模型和 TCP/IP 的原理。

【预备知识】

一、OSI 参考模型

OSI/RM（Open System Interconnection，开放系统互联参考模型）是由 ISO（国际标准组织）创建的一个有助于开放和理解计算机的通信模型。OSI 参考模型作为一套规范的标准，在网络世界中得到广泛的应用。它把网络结构分为 7 个层次，从低到高依次是物理层、数据链路层、网络层、传输层、会话层、表示层和应用层。

在 OSI 模型中，每一层具有独立的功能，上下层之间能够完成网络信息的交互。每一层主要的功能如表 1-4-1 所示。

表 1-4-1　　　　　　　　　　OSI 参考模型各层的主要功能

OSI 模型各层的名称	主要功能
物理层	传输比特流
数据链路层	提供介质访问、链路管理等
网络层	路由选择、拥塞控制和网际互联
传输层	为上层协议提供端到端的可靠和透明的数据传输服务
会话层	负责建立、管理和终止应用程序之间的会话
表示层	处理数据格式、数据加密等
应用层	提供应用程序接口

信息从发送方到接收方传递的路线为发送方 AP_A→应用层→表示层→……→物理层→物理媒体（如双绞线、光纤等），信息经过中间结点的中继后→接收方的物理层→……→应用层→AP_B。

信息在 OSI 模型各层流动，其数据结构不断发生着变化。应用进程 AP_A 先将数据交给 OSI 模型的应用层，应用层加上自己的控制信息就变成下一层（表示层）的数据单元。表示层收到这些数据单元后加上本层的控制信息交给下一层（会话层），变成下一层的数据单元……依次类推，而最底层物理层只是比特流的传送，不加任何控制信息，由物理媒体传送到接收方，接收方就将信息从第一层（物理层）依次传送到第七层（应用层）。每一层根据控制信息进行必要的操作，并去除附加在信息上的相应控制信息内容，最后把应用进程 AP_A 的原始数据正确地交给目的站点的应用进程 AP_B。数据结构变化情况如图 1-4-1 所示。

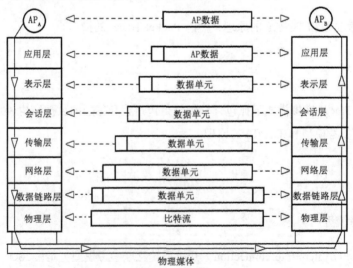

图 1-4-1　OSI 模型中的数据流

二、TCP/IP

TCP/IP（Transmission Control Protocol/Internet Protocol）是由美国国防部高级研究计划局研究创立的，它由两个主要的协议 TCP 和 IP 而得名。TCP/IP 协议是一种工业标准协议组，是为在大型的、由异构系统组成的网络通信而设计的一种开放的协议组。TCP/IP 组中的协议分为四个层次：网络接口层、网络层、传输层和应用层。

TCP/IP 开发工作比 OSI 模型标准的制定还要早，与 OSI 模型体系结构存在着一定的对应关系。对应关系如表 1-4-2 所示。

表 1-4-2　　　　　　　　　　OSI 和 TCP/IP 的层次关系

OSI 参考模型	TCP/IP 协议
应用层	应用层
表示层	
会话层	
传输层	传输层
网络层	网络层
数据链路层	网络接口层
物理层	

三、TCP/IP 的各层功能及主要协议

（1）应用层。

定义了 TCP/IP 应用协议以及处理应用程序的具体细节，应用层是应用程序进入网络的通道。在应用层有许多 TCP/IP 工具和服务，如 HTTP、FTP、Telnet、DNS 等。

（2）传输层。

提供主机之间的通信会话管理，并且定义了传输数据时的服务级别和连接状态。工作在这一层的传输协议在计算机之间提供会话连接的建立，并负责将应用层的数据向 IP 层传递或将 IP 层的数据向应用层传递。传输层具有两个核心协议，用于提供数据传输的方法。

① TCP 传输控制协议。

传输控制协议（TCP，Transport Control Protocol）提供可靠的、面向连接的数据包传递服务。传输控制协议可以确保 IP 数据包的成功传递，对程序发送的大块数据进行分段和重组，可以确保正确排序以及按顺序传递分段的数据。

② UDP 用户数据包协议。

UDP 提供快速的、无连接的、不可靠的数据包传输服务。不同于 TCP，UDP 提供"尽最大努力传递"的无连接数据包服务，不返回确认信息，不保证数据包的有序性以及不提供出错包的重传机制。

（3）网络层。

网络层将上层传下来的数据装入 IP 数据包，封装 IP 包头，包含用于在主机间以及经过在网络转发数据时需要用到的源和目标的地址信息，以实现 IP 数据报的路由。网络层主要包含以下几个协议。

① IP 协议（网际协议）。

IP 是一种无连接的、不可靠的协议，主要用于编址数据包并负责路由数据包。可以将 IP 看作 TCP/IP 协议组中的邮局，从事数据包的存储和转发工作。来自传输层的 TCP 或 UDP 数据包以及来自下层（网络接口层）的数据包在 IP 层中进行地址的标识和路由，最终送往目的地。

每个数据包中都封装有源 IP 地址和目标 IP 地址。如果源 IP 地址和目标 IP 地址在相同的网段，那么就直接从源发送到目标。如果不是，那么 IP 就使用一个适当的路由进行发送。IP 还将定义数据包的 TTL 值（生命值），TTL 值决定了数据包在网络上的最长传输时间，超时将被丢弃。

② ARP（地址解析协议）。

在以太网协议中规定，同一局域网中的一台主机要和另一台主机进行直接通信，必须要知道目标主机的 MAC 地址。而在 TCP/IP 协议栈中，网络层和传输层只关心目标主机的 IP 地址。这就导致在以太网中使用 IP 协议时，数据链路层的以太网协议接到上层 IP 协议提供的数据中，只包含目的主机的 IP 地址。于是需要一种方法，根据目的主机的 IP 地址，获得其 MAC 地址。这就是 ARP 协议要做的事情。所谓地址解析（address resolution）就是指主机在发送帧前将目标 IP 地址转换成目标 MAC 地址的过程。

另外，当发送主机和目的主机不在同一个局域网中时，即便知道目的主机的 MAC 地址，两者也不能直接通信，必须经过路由转发才可以。所以此时，发送主机通过 ARP 协议获得的将不是目的主机的真实 MAC 地址，而是一台可以通往局域网外的路由器的某个端口的 MAC 地址。于是此后发送主机发往目的主机的所有帧，都将发往该路由器，再通过它向外发送。这种情况称为 ARP 代理。

③ RARP（反向地址解析协议）。

ARP 协议是根据 IP 地址找到对应的 MAC 地址，而 RARP 则是根据 MAC 地址找到对应的 IP 地址，所以称之为"反向 ARP"。具有本地磁盘的系统引导时，一般是从磁盘上的配置文件中读取 IP 地址，然后即可直接用 ARP 协议找出与其对应的主机 MAC 地址。但是无盘机，如 X 终端或无盘工作站，启动时是通过 MAC 地址来寻址的，这时就需要通过 RARP 协议获取 IP 地址。

④ ICMP（INERNET 控制信息协议）。

为数据包的传输提供诊断功能以及错误报告，基于 IP 通信的计算机或路由器通过 ICMP 就能够检测到错误并交换控制和状态信息。我们经常使用的诊断命令 ping 和 traceroute 使用的就是 ICMP 协议。

【实训拓扑】

网络拓扑结构图如图 1-4-2 所示。

图 1-4-2　网络拓扑结构图

【实训设备】

计算机 2 台、交叉线一根。

【实训步骤】

步骤 1　在模拟方式中构建 OSI 环境。

表 1-4-3　（1）运行 Packet Tracer 5.3，按拓扑图搭建网络，为 PC1 和 Server1 配置 IP 地址，如表 1-4-3 所示。

设备名称	接口	IP	子网掩码
PC1	网卡	192.168.1.1	255.255.255.0
Server1	网卡	192.168.1.2	255.255.255.0

在 Packet Tracer 中配置计算机 IP 地址的方法如图 1-4-3 和图 1-4-4 所示。

图 1-4-3　PC1 的 IP 配置　　　　　　　图 1-4-4　Server1 的 IP 配置

（2）单击模拟模式按钮进入模拟环境，如图 1-4-5 所示。

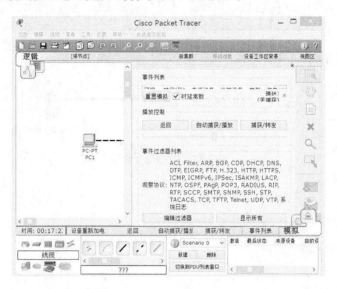

图 1-4-5　模拟方式

（3）单击"编辑过滤器"按钮，只保留 ARP、HTTP 两个选项，如图 1-4-6 所示。

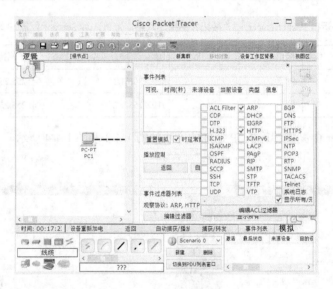

图 1-4-6　编辑过滤器

（4）单击 PC1，选择"桌面"选项卡，单击 Web 浏览器，在 URL 地址栏中输入"192.168.1.2"，再单击右边的"跳转"按钮，如图 1-4-7 所示。

项目1 网络基础

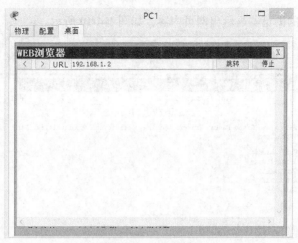

图1-4-7 PC1浏览器

（5）连续单击模拟区中的"捕获/转发"按钮，数据包以信封的形式在PC1和Server1之间流动。直至看到数据包回到PC1并呈现绿色对号为止，如图1-4-8所示。

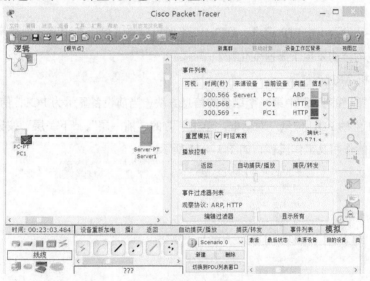

图1-4-8 捕获/转发数据包

（6）在模拟区"事件列表"中出现类型为ARP和HTTP的信息，详细列出每一数据包的来源和当前在哪个设备上，如图1-4-9所示。

图1-4-9 事件列表

21

（7）最终，PC1 的浏览器显示出网页内容，如图 1-4-10 所示。

图 1-4-10　PC1 的浏览器

思考：PC1 从发送出 HTTP 请求，到 Server1 做出应答，最后在浏览器中显示出网页内容。这期间，数据包的结构发生了哪些变化呢？

步骤 2　分析 ARP。

（1）单击"事件列表"中第一个绿色的信息方块，当前设备显示为 PC1，信息类型为 ARP。弹出"设备 PC1 上的 PDU 信息"对话框。单击下方的"前一层"，"下一层"按钮，会显示出在 OSI 相应层的功能作用及工作流程，如图 1-4-11 所示。

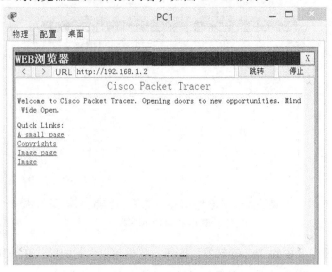

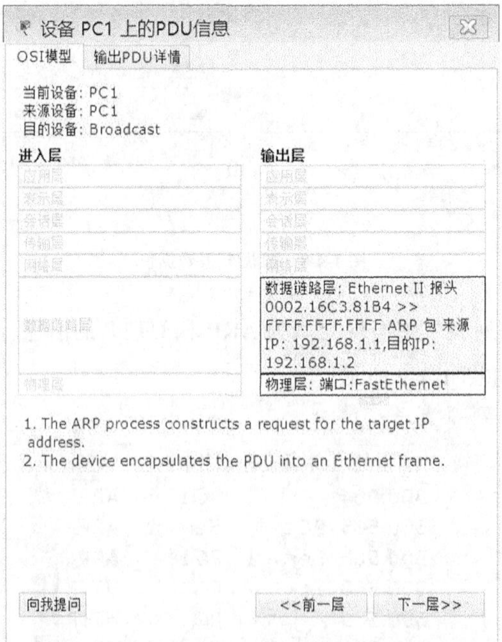

图 1-4-11　PC1 发送的 PDU 信息

该对话框显示：PC1 构建一个 ARP 请求，协议类型为 0x1，要求解析目标 IP 是 192.168.1.2 的 MAC 地址，因为目的 IP 的 MAC 不详，设置为 "0000.0000.0000"。ARP 协议工作在 OSI 参考模型第三层网络层，ARP 请求的数据单元传递到数据链路层，被加上数据链路层的控制信息。在数据链路层，数据的单位叫"帧"。PC1 的 MAC 为"0002.16C3.8184"，目的 MAC 为"FFFF.FFFF.FFFF"表示是一个广播帧。该帧经过物理层的网卡发出，在本地以太网内广播，本地以太网中的计算机都会接收到。我们单击"输出 PDU 详情"，可以观察到帧的结构及包含的 ARP 请求信息，如图 1-4-12 所示。

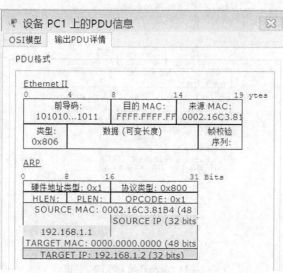

图 1-4-12　PC1 输出的 PDU 详情

（2）单击"事件列表"中第二个绿色的信息方块，当前设备显示为 Server1，信息类型为 ARP。弹出"设备 Server1 上的 PDU 信息"对话框，如图 1-4-13 所示。

图 1-4-13　Server1 上的 PDU 信息

在"进入层"的物理层，网卡接收到广播帧，将其上传给数据链路层。数据链路层去除本层的控制信息，因为 ARP 请求解析的目标 IP192.168.1.2 和自己的一致，ARP 处理该请求帧，将接收的信息更新 ARP 表。然后，ARP 进程用接收的 MAC 地址应答请求，在"输出层"的数据链路层进入封装，源 MAC 为自己的网卡地址，目的 MAC 为接收帧的源 MAC，即 PC1 的网卡地址。ARP 应答包作为数据单元被封装在应答的数据帧内部，协议类型变为 0x2。该帧经过物理层的网卡发出。同样，我们单击"输出 PDU 详情"，可以看到数据结构的详细情况，如图 1-4-14 所示。

图 1-4-14　Server1 输出的 PDU 详情

（3）单击"事件列表"中第三个绿色的信息方块，弹出"设备 PC1 上的 PDU 信息"对话框，可以看到在物理层上网卡接收到比特流。单击"下一层"按钮，在工作流程区看到由于接收帧的目的 MAC 和接收网卡 MAC 一致，所以数据链路层拆封，由 ARP 进程处理该应答帧，将接收到的信息更新到 ARP 地址表中，如图 1-4-15 所示。

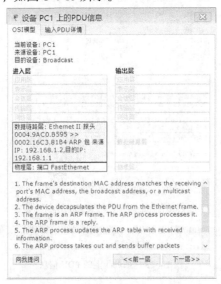

图 1-4-15　PC1 接收的 PDU 信息

步骤 3 分析 HTTP 信息的传递。

（1）单击"事件列表"中第一个类型为 HTTP 的信息方块，弹出"设备 PC1 上的 PDU 信息"对话框。此时会发现 HTTP 客户端在应用层发出一个 HTTP 请求，如图 1-4-16 所示。

图 1-4-16　应用层封装 HTTP 请求

单击"输出 PDU 详情"选项卡，查看 HTTP 请求的 PDU 结构，如图 1-4-17 所示。

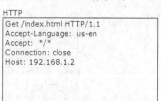

图 1-4-17　HTTP 请求信息

（2）单击"OSI 模型"选项卡下方的"下一层"按钮。在传输层发送报文：序列号为 1，ACK（确认）号为 1，数据长度为 99。如图 1-4-18 所示。

图 1-4-18　传输层进行封装

查看输出 PDU 详情，发现传输层采用 TCP 协议将上层发送的数据单元进行封装（HTTP 请求等信息被包装在数据中），并加上自己的控制信息——TCP 包头：源端口号、目的端口、序列号、确认序列号、首部长度、标志位、窗口大小、校验和、紧急指针等。如图 1-4-19 所示。

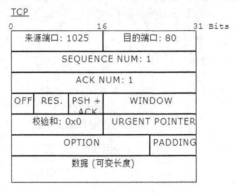

图 1-4-19　TCP 报文

（3）单击"下一层"按钮，因为目的 IP 在同一子网，将目的 IP 设置为下一跳，如图 1-4-20 所示。

图 1-4-20　网络层进行封装

查看输出 PDU 详情，发现网络层采用 IP 协议将上层发送的数据单元进行封装，并加上自己的 IP 协议包头：版本、包头长度 IHL、服务类型、包长度 TL、标识、标记、分段偏移、生存时间 TTL、协议 PRO、校验和、源 IP、目的 IP 等。上层发送的数据单元被封装在可变长度的数据中，如图 1-4-21 所示。

项目1 网络基础

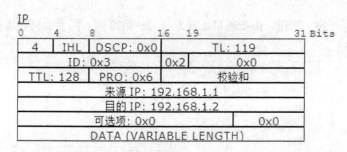

图 1-4-21　IP 数据包

（4）单击"下一层"按钮，下一跳 IP 地址是一个单播地址，ARP 进程查找 ARP 表。下一跳 IP 地址在 ARP 表中，ARP 进程设置目的 MAC 为表中查找到的一项值。网卡将该 PDU 封装进一个以太网帧，如图 1-4-22 所示。

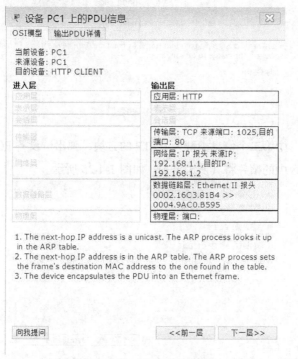

图 1-4-22　数据链路层进行封装

查看输出 PDU 详情，发现数据链路层将上层发送的数据单元进行了封装，并加上自己的帧头（6 字节的目的 MAC 地址和 6 字节的源 MAC 地址、2 字节的类型字段），还在帧尾加上 4 字节的帧校验序列，如图 1-4-23 所示。

图 1-4-23　数据帧结构

27

（5）单击"下一层"按钮，由于端口正在发送另一数据帧，网卡会将帧缓存下来稍后再发送，如图 1-4-24 所示。

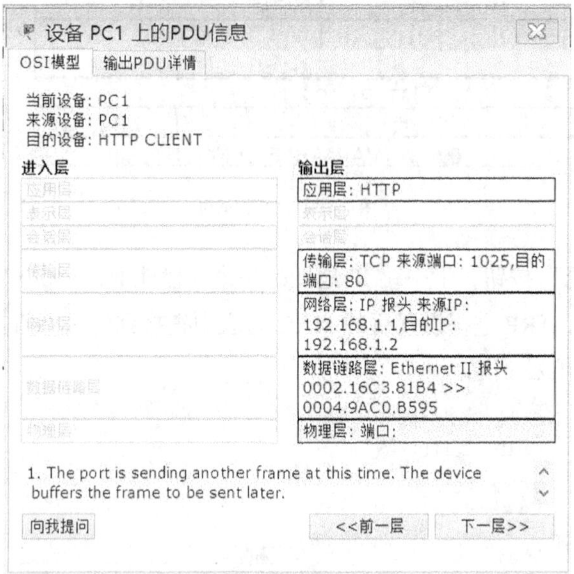

图 1-4-24 物理层

单击"事件列表"中第二个信息类型为 HTTP 的方块，弹出"设备 PC1 上的 PDU 信息"对话框。网卡取出缓冲区中的帧并将其发送出去，如图 1-4-25 所示。

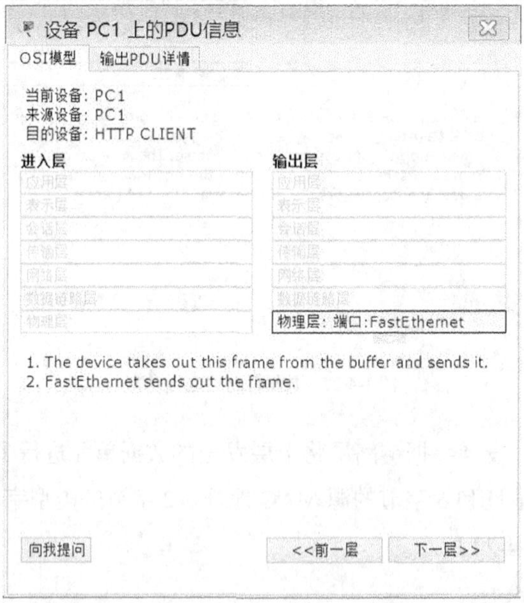

图 1-4-25 PC1 在物理层上发送帧

由以上过程可知，物理层对数据不做任何处理，只是发送和接收比特流。

（6）单击"事件列表"中第三个信息类型为 HTTP 的方块，弹出"设备 Server1 上的 PDU 信息"对话框。显示物理层接收到帧，如图 1-4-26 所示。

项目 1　网络基础

图 1-4-26　Server1 物理层接收到帧

（7）单击"下一层"按钮，数据上传给数据链路层，帧的目的 MAC 匹配接收端口的 MAC 地址，网卡从帧中解封出 PDU 上传给网络层，如图 1-4-27 所示。

图 1-4-27　数据链路层拆封帧

29

（8）单击"下一层"按钮，包中的目的 IP 和本网卡 IP 地址一致，网卡对包进行拆解并上传给传输层，如图 1-4-28 所示。

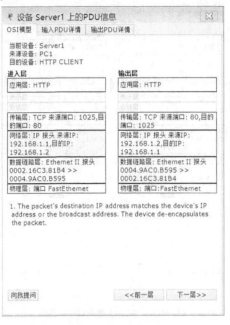

图 1-4-28　网络层拆封包

（9）单击"下一层"按钮。在传输层上，网卡接收到来自 192.168.1.1（PC1）端口号 1025 的 TCP PUSH+ACK 报文，报文信息显示序列号 1、ACK 号 1，数据长度 99。该报文包含所期望的序列号，TCP 处理负载的数据,然后重新封装所有数据并交给更高层，如图 1-4-29 所示。

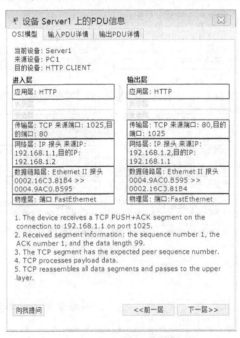

图 1-4-29　Server1 传输层

（10）单击"下一层"按钮，应用层上显示服务器接收到一个 HTTP 请求，如图 1-4-30 所示。

图 1-4-30　服务器接收到 HTTP 请求

（11）单击"输入 PDU 详情"选项卡，看到接收的 HTTP 请求与发送的内容完全一样，如图 1-4-31 所示。

至此，PC1 发送的 HTTP 请求经过从应用层到数据链路层的层层封装，在物理层转化为比特流，通过物理媒体传输到 Server1；Server1 物理层接收到信息后上传，经过数据链路层到应用层的层层抓封，还原出原始的信息。

（12）单击"下一层"按钮，Server1 的应用层显示服务层发回一个 HTTP 应答。单击"输出 PDU 详情"，HTTP 应答内容如图 1-4-32 所示。

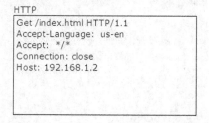 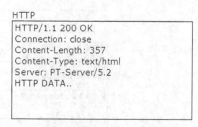

图 1-4-31　Server1 接收的 HTTP 请求内容　　图 1-4-32　HTTP 应答内容

（13）连续单击"下一层"按钮，显示信息传送过程为从应用层向下，各层封装上层信息，并加上自己的控制信息。其中，传输层的 ACK 号为 100 表示自己成功接收 PC1 发送的 99 字节数据，希望对方接着从第 100 字节发送数据。源端口和目的端口、源 IP 和目的 IP、源 MAC 和目的 MAC 正好和接收的相反。如图 1-4-33 所示。

图 1-4-33　Server1 发回 HTTP 应答

（14）单击"事件列表"中第四个信息类型为 HTTP 的方块，弹出"设备 PC1 上的 PDU 信息"对话框。显示信息传送过程如下：物理层接收到帧；数据上传给数据链路层，帧的目的 MAC 匹配接收端口的 MAC 地址，网卡从帧中解封出 PDU 上传给网络层；包中的目的 IP 和本网卡 IP 地址一致，网卡对包进行拆解上传给传输层；传输层上，网卡接收到来自 192.168.1.2（PC1）端口号 80 的 TCP PUSH+ACK 报文，报文信息显示序列号 1、ACK 号 100 和数据长度 459。该报文包含所期望的序列号，TCP 协议处理负载的数据，然后重新封装所有数据并交给更高层。如图 1-4-34 所示。

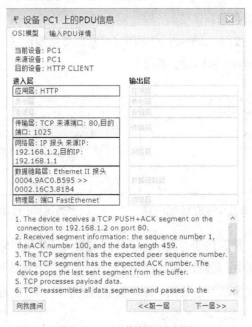

图 1-4-34　PC1 传输层进行拆封

（15）单击"下一层"按钮，应用层上显示 HTTP 客户端接收到 HTTP 应答，它在浏览器中显示网页内容，如图 1-4-35 所示。

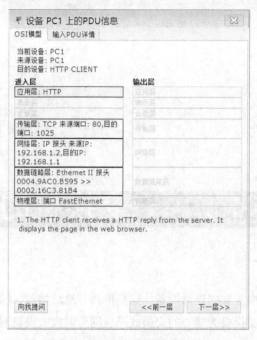

图 1-4-35　PC1 应用层接收到 HTTP 应答

从 PC1 发出 HTTP 请求到接收到 Server1 的 HTTP 应答，这个数据传送过程相当漫长，且数据结构变化也复杂，但由于 OSI 模型和 TCP/IP 对用户是屏蔽的，所以用户根本体会不到其中的艰辛。就像我们只看到有的人能够取得好成绩，却不知道他们私下付出了多少血汗一样。

项目 2
交换机的基本配置

　　交换机是局域网中重要的连通设备。交换机具有地址学习，帧的转发、过滤以及环路避免功能。采取以交换机为核心的交换式局域网可以大大提高网络性能。

　　通过本项目的学习，我们可以掌握交换机的基本配置命令、能够配置交换机端口安全、理解生成树的原理和基本配置以及学会划分 VLAN 实现部门间的隔离等。

任务 2.1　交换机配置方式及基本操作

【实训目的】

通过对交换机设备的几种配置手段、配置模式和基本配置命令的认识，掌握交换机的基本使用能力。

【实训任务】

1. 掌握交换机命令行各种操作模式的区别，模式之间的切换。
2. 熟练掌握交换机配置的基本命令。

【预备知识】

一、什么是交换机

交换机（Switch）是一种在通信系统中完成信息交换功能的设备。交换机基于数据帧中的目的 MAC 地址转发信息，它为接入的任意两个网络节点提供独享的电信号通路。我们常见的交换机是以太网交换机。从 OSI 体系结构来看，传统的交换机属于 OSI 的第二层数据链路层设备。目前广泛应用的是三层交换机，而四层交换机也逐步得到应用。

二、交换机的主要功能

交换机的主要功能包括学习功能、转发过滤和环路避免。

1. 学习功能：当交换机开机初始化时，它的 MAC 地址是空白的。当它从某个端口接收到一个数据包时，首先读取帧头中的源 MAC 地址，学习到源 MAC 地址的设备是连在哪个端口上的，并将其添加到 MAC 地址表中；然后再去读取帧头中的目的 MAC 地址，并在 MAC 地址表中查找相应的端口；如果 MAC 表中有和目的 MAC 地址对应的端口，就把数据包直接复制到该端口上；如果 MAC 地址表中找不到目的 MAC 地址对应的端口，就向除了接收到数据包的源端口之外的所有端口进行广播，当目标设备对源机器回应时，交换机又可以学习到目的机器的 MAC 地址和哪个端口对应，并将其添加到自己维护的 MAC 地址表中。当相同的源设备和目的设备再次传送数据时就可以直接将数据包转发到目的端口，而不再进行广播了。

通过不断重复以上学习过程，交换机完全掌握了全网的 MAC 地址信息，在缓存中建立和维护一张 MAC 地址表。

2. 转发过滤：当一个数据帧的目的 MAC 地址在 MAC 地址表中有映射时，它就会被转发到连接目的设备的端口而不是所有端口（如果该数据帧是广播帧或者组播帧则转发至所有端口）。

3. 环路避免：当交换机包括一个冗余回路时，以太网交换机通过生成树协议避免回路的产生，同时允许存在后备路径。

三、交换机工作模式

可网管交换机工作模式有三种：用户模式、特权模式和配置模式。

工作模式切换的命令归纳为图 2-1-1 所示。

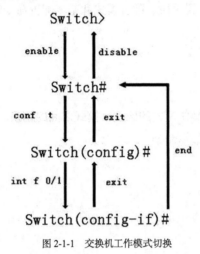

图 2-1-1　交换机工作模式切换

【实训拓扑】

网络拓扑结构图如图 2-1-2 所示。

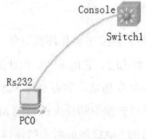

图 2-1-2　网络拓扑结构图

【实训设备】

PC 1 台、交换机 1 台、DB-9 适配器 1 个、反转线 1 根。

【实训步骤】

步骤 1　构建本地配置环境（Packet Tracer 配置环境在后续章节中讲解）。

（1）将随交换机附带的配置线一端连接计算机的串口，另一端连接交换机的 Console 口。如果用自己制作的反转线连接，则反转线的一端通过 DB-9 适配器连接到 PC 机的串口（即 COM 口），反转线缆的另一端连接交换机的 Console 口。

（2）单击"开始"按钮，在"程序"菜单的"附件"选项中单击"超级终端"。在弹出的对话框中输入城市区号，如图 2-1-3 所示。

图 2-1-3 位置信息

（3）在"名称"文本框中键入需新建的超级终端连接项名称，然后单击"确定"按钮，如图 2-1-4 所示。

图 2-1-4 新建连接

（4）弹出"连接到"对话框，在"连接时使用"下拉列表框中选择与路由器相连的计算机的串口，如图 2-1-5 所示。

图 2-1-5 "连接到"对话框

(5)单击"确定"按钮,弹出 COM1 属性对话框,如图 2-1-6 所示。

图 2-1-6 COM1 属性对话框

(6)在 COM1 属性中单击"还原为默认值"按钮,再单击"确定"按钮。

(7)按 Enter 键,出现交换机标识符:switch>。

步骤 2 交换机配置模式切换和基本命令。

(1)配置模式切换。

```
Switch>                        //普通用户模式
Switch>enable                  //进入特权用户模式
Switch#                        //特权用户模式
Switch#configure  terminal     //进入全局配置模式
Switch(config)#                //全局配置模式
Switch(config)#interface  f0/1 //进入接口配置模式
```

```
Switch(config-if)#              // 接口配置模式
Switch(config-if)#exit          //返回上一级模式
Switch(config)#
Switch(config)#vlan    10       //进入 VLAN 配置模式
Switch(config-vlan)#            //VLAN 配置模式
Switch(config-vlan)#exit        //返回上一级模式
Switch(config)#exit             //返回上一级模式
Switch#
Switch#disable                  //退出特权模式
Switch>
Switch(config-if)#end           //直接返回特权模式
Switch#
```

（2）修改交换机的名字(Hostname)。

```
Switch# configure   terminal
Switch(config)#hostname  switch1    //修改交换机名字为"switch1"
```

（3）设置交换机端口参数。

```
switch1(config)#interface   fastEthernet   0/1
switch1(config-if)#speed    100     //配置端口速率为 100Mbit/s
switch1(config-if)#duplex   half    //配置为半双工模式
switch1(config-if)#no   shutdown    //开启端口
switch1(config-if)#
```

（4）配置交换机管理地址。

```
switch1(config-if)#exit
switch1(config)#interface   vlan   1    //打开交换机的管理 VLAN
switch1(config-if)#ip address   192.168.1.1   255.255.255.0   //配置管理地址为 192.168.1.1/24
switch1(config-if)#no    shutdown
switch1(config-if)#exit
switch1(config)#
```

（5）管理交换机配置信息。

```
switch1#copy    running-config    startup-config //将当前运行的参数保存到启动文件中
switch1#write    //将当前配置信息保存到 flash 中用于系统初始化
switch1#delete   flash:config.text   //永久删除 flash 中的配置文件
```

任务 2.2　设置交换机密码

【实训目的】

掌握交换机控制台密码、远程登录密码和特权密码的设置，加强安全管理。

【实训任务】

1. 设置交换机特权密码。
2. 设置交换机控制台密码。
3. 设置交换机远程登陆密码。

【预备知识】

一、设置交换机特权密码

用户在交换机用户模式下输入 enable 命令，直接进入特权模式，就会拥有管理该交换机的所有权力，这样给网络设备管理带来了巨大的风险。设置交换机特权密码以后，在从用户模式下进入特权模式时，需要用户输入正确的口令，只有验证通过才能进入特权模式，增加了交换机的安全性。

特权密码又分加密密码和明文密码两种。加密密码优先级高于明文密码。如果同时设置了加密密码和明文密码，则明文密码失效。此外，加密密码无法正常显示，而明文密码我们可以在配置文件中看到。

在全局配置模式下，设置交换机特权密码的命令如下。

设置加密密码：switch (config)#enable secret　密码

设置明文密码：switch (config)#enable password　密码

二、设置交换机控制台密码

交换机前面通常都有一个 console 接口，我们用交换机附带的配置线一端连接计算机的串口，另一端接到交换机的 console 接口，然后通过超级终端对交换机进行带外管理。

在全局配置模式下，设置控制台密码的命令如下。

switch(config)#line console 0

switch (config-line)#password　密码

switch (config-line)#login

三、设置交换机 Telnet 密码

设置交换机 Telnet 密码后,我们就可以通过计算机自带的 telnet 工具远程登陆交换机,给网络设备的管理带来灵活性。

在全局配置模式下,设置 Telnet 密码的命令如下。

switch(config)#line vty 0 4
switch (config-line)#password 密码
switch (config-line)#login

【实训拓扑】

网络拓扑结构图如图 2-2-1 所示。

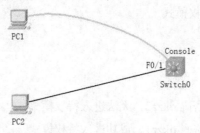

图 2-2-1 网络拓扑结构图

【实训设备】

交换机 1 台、计算机 2 台

【实训步骤】

步骤 1 设置特权密码。

(1)我们在 PC1 上通过超级终端连入交换机,设置如图 2-2-2 所示。

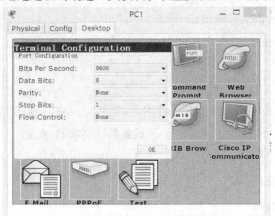

图 2-2-2 超级终端

单击"OK"按钮,进入交换机用户模式。

Switch>

Switch> enable　　//由于尚未设置特权密码,可以直接进入到特权模式

Switch#

(2)配置交换机特权密码。

Switch#configure　Terminal　　//进入全局配置模式

Switch(config)#hostname　switch0　　//设置交换机名字为 switch0

switch0(config)#enable　secret　qkzz1　　//设置加密密码为 qkzz1

switch0(config)#enable　password　qkzz2　　//设置明文密码为 qkzz2

switch0(config)#exit　　//返回特权模式

switch0#

switch0#disable　//退出特权模式

switch0>

(3)测试特权密码。

switch0>enable

Password:　　//输入明文密码 qkzz2,无法进入特权模式

Password:　　//输入加密密码 qkzz1,成功进入特权模式

switch0#

经过测试,明文密码和加密密码同时存在时,明文密码失效,表明加密密码优先级高于明文密码。

(4)显示配置文件。

switch0#show　running-config

Building configuration---

Current configuration : 1115 bytes

version 12-2

no service timestamps log datetime msec

no service timestamps debug datetime msec

no service password-encryption

hostname switch0

enable secret 5 1mERr$dw4lAQEFWmTlNsUsSIQ5p-　　//加密密码无法正常显示

enable password qkzz2　　　　　　　//设置的明文密码 qkzz2

步骤 2　设置控制台密码。

(1)设置 console 密码。

switch0#configure terminal

Enter configuration commands, one per line.   End with CNTL/Z.
switch0(config)#line console 0   //进入控制台口状态
switch0(config-line)#password   qkzz3   //设置登录口令为 qkzz3
switch0(config-line)#login   //允许登录
switch0(config-line)#end   //直接切换到特权模式
switch0#write   //保存配置信息
Building configuration---
[OK]
switch0#
（2）显示配置文件。
switch0#show   running-config
------
line   con   0
 password   qkzz3   //设置的控制台密码 qkzz3
 login

（3）测试控制台密码。
switch0#reload   //重启交换机

Loading "flash:/c3560-advipservicesk9-mz-122-37-SE1.bin"---
[OK]

Press RETURN to get started!
User Access Verification
Password: //输入控制台密码 qkzz3
switch0>
步骤 3 设置 Telnet 密码。
（1）配置 Telnet 密码。
switch0>enable
Password: //输入加密特权密码 qkzz1
switch0#configure   terminal
Enter configuration commands, one per line.   End with CNTL/Z.
switch0(config)#line   vty   0 4   //进入虚拟终端配置模式
switch0(config-line)#password   qkzz4   //设置登录口令 qkzz4
switch0(config-line)#login   //允许登录
switch0(config-line)#exit   //返回上一层

（2）设置交换机管理 vlan 的 IP 地址。

switch0(config)#interface　　vlan　　1　　　//进入交换机 vlan11 虚拟口

switch0(config-if)#ip　　address　192.168.1.254　　255.255.255.0　//给 vlan1 设置 IP 地址

switch0(config-if)#no　　shutdown　　//开启 vlan1

switch0(config-if)#end　　　　　　　　//直接返回特权模式

（3）显示配置文件。

switch0#show　running-config

------

interface Vlan1

　ip address 192.168.1.254 255.255.255.0

line con 0

　password qkzz3

　login

line vty 0 4

　password qkzz4

　login

（4）测试 Telnet 密码。

我们首先配置 PC2 的 IP 地址，如 192.168.1.10,子网掩码为 255.255.255.0。PC2 的 IP 必须与交换机管理 vlan 在同一网段。

然后再在 PC2 上运行 Telnet：

　PC>telnet　　192.168.1.254

Trying 192.168.1.254 ---Open

User Access Verification

Password:　　//输入远程登录 Telnet 密码 qkzz4

switch0>　　　//进入交换机用户模式

switch0>enable

Password:　　//输入加密特权密码 qkzz1

switch0#　　　//进入交换机特权模式

控制台密码、Telnet 密码是用户以不同的方式登录交换机的时候被要求输入的密码，不同登录方式的密码可能被设置的不一样；而明文特权密码和加密特权密码是用户在用户模式下输入 enable 进入特权模式时被要求输入的。加密特权密码优先级高于明文特权密码。如果同时设置了加密特权密码和明文特权密码，则明文特权密码失效。此外，加密特权密码无法正常显示，而明文特权密码我们可以在配置文件中看到。

任务 2.3 划分 VLAN，实现部门间隔离

【实训目的】

理解 IP 地址和子网掩码的作用，掌握划分子网的方法，并能熟练地在交换机上创建 VLAN。

【实训任务】

1. 根据各部门的计算机数量划分 VLAN。
2. 在二层交换机上构建 VLAN，并测试相同 VLAN 中计算机的连通性。
3. 测试不同 VLAN 中计算机的连通性。

【预备知识】

一、VLAN 的功能和作用

VLAN 即虚拟局域网（Virtual Local Area Network 的缩写），是一种通过将局域网内的设备逻辑地划分成一个个网络的技术。

交换机有一个默认的 VLAN1，管控着所有的端口，是一个很大的广播域。而划分出的一个 VLAN 就是一个小的广播域，单播帧和广播帧都在 VLAN 内部扩散转发，不会进入到其他 VLAN 中。划分 VLAN 提高了网络安全性，控制了广播带来的带宽消耗，也增加了网络连接的灵活性。

我们最常用的是基于交换机端口的 VLAN 划分。

二、IP 地址和子网掩码

一个 IP 地址用来标识网络中的一个通信实体，如一台主机或路由器的某一个端口。

IP 地址是一个 32 位的二进制数，常用点分十进制的方法来表示。IP 地址由网络号和主机号这两部分组成。网络号用于标识该地址从属哪个网络；主机号用于指明该网络上的某个特定主机。

IP 地址分为 5 类：A 类、B 类、C 类、D 类和 E 类。其中 D 类用于组播，E 类为保留地址，没有分配使用。

子网掩码的长度也是 32 位的二进制数，左边是网络位，用"1"表示，1 的位数等于网络位的长度；右边是主机位，用"0"表示，0 的位数等于主机位的长度。子网掩码与 ip 地址做"与"运算可以取出网络号。

A 类网络首字节取值范围：1～126   子网掩码：255.0.0.0
B 类网络首字节取值范围：128～191   子网掩码：255.255.0.0
C 类网络首字节取值范围：192～223   子网掩码：255.255.255.0

三、划分子网的方法

划分子网时，原 IP 地址的网络位保持不变，将主机位的最高几位借出来划分子网。子网划分后的 ip 地址由三部分组成：网络号+子网号+子网主机号。借出划分子网的主机位对应的子网掩码位由"0"变为"1"。

如何确定要借出的主机位数呢？一是根据要划分的子网数量；二是根据每个子网需要的最大主机数。两个因素要综合考虑，同时满足方可。

确定子网掩码的方法如下。
1. 将子网所需的最大主机数转化为二进制数，假设二进制位数为 N。
2. 将子网掩码的 32 个二进制位全部置为"1"。然后将其后面的 N 位改为"0"。
3. 计算出子网掩码。

【实训拓扑】

网络拓扑结构图如图 2-3-1 所示。

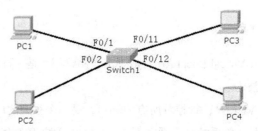

图 2-3-1 网络拓扑结构图

【实训设备】

交换机 1 台、计算机 4 台。

【实训步骤】

步骤 1 划分 VLAN。

青岛开发区职专东校区有两个专业部：信息部和社会服务部。每个专业部教师数量不超过 60 人。东校区使用私有网段 192.168.10.0。要求将两部门划入不同的 VLAN，实现部门间的隔离。

（1）确定子网掩码。

① 将子网所需的最大主机数转化为二进制数。十进制的 60 转化为二进制为 111100,所以要保留 6 位主机位，可借出主机位的前 2 位划分子网。

② 将子网掩码的 32 个二进制位全部置为"1"，即"11111111 11111111 11111111 11111111"。然后，再将后面的 N 位改为"0",结果为"11111111 11111111 11111111 11000000",转化成点分十进制就是 255.255.255.192。

（2）计算出各 VLAN 的网络地址、广播地址，得出可分配的 IP 范围。

192.168.10.0/24 是 C 类网，前三字节网络号无须改变，将后一字节主机号变为八位二进制形式。由二进制子网掩码"11111111　11111111　11111111　11000000"可知，原主机号借出前 2 位划分子网，可能的状态如表 2-3-1 所示。

表 2-3-1

网络号	子网号	主机号
192.168.10.	00	000000……111111
192.168.10.	01	000000……111111
192.168.10.	10	000000……111111
192.168.10.	11	000000……111111

因为主机号全"0"为网络地址，主机号全"1"为广播地址，得出以下 4 点结论。

第一个 VLAN 的网络地址是：192.168.10.00000000，点分十进制形式为 192.168.10.0。广播地址是：192.168.10.00111111，点分十进制形式为 192.168.10.63。可分配的 IP 范围为 192.168.10.1～192.168.10.62。

第二个 VLAN 的网络地址是：192.168.10.01000000，点分十进制形式为 192.168.10.64。广播地址是：192.168.10.01111111，点分十进制形式为 192.168.10.127。可分配的 IP 范围为 192.168.10.65～192.168.10.126。

第三个 VLAN 的网络地址是：192.168.10.10000000，点分十进制形式为 192.168.10.128。广播地址是：192.168.10.10111111，点分十进制形式为 192.168.10.191。可分配的 IP 范围为 192.168.10.129～192.168.10.190。

第四个 VLAN 的网络地址是：192.168.10.11000000，点分十进制形式为 192.168.10.192。广播地址是：192.168.10.11111111，点分十进制形式为 192.168.10.255。可分配的 IP 范围为 192.168.10.193～192.168.10.254。

划分结果如表 2-3-2 所示。

表 2-3-2　　　　　　　　　　　　　　划分 VLAN

部门名称	网络地址	广播地址	可分配 IP 范围
信息部	192.168.10.0	192.168.10.63	192.168.10.1 — 192.168.10.62
社服部	192.168.10.64	192.168.10.127	192.168.10.65 — 192.168.10.126

续表

部门名称	网络地址	广播地址	可分配 IP 范围
待分配	192.168.10.128	192.168.10.191	192.168.10.129 — 192.168.10.190
待分配	192.168.10.192	192.168.10.255	192.168.10.193 — 192.168.10.254

（3）本实验在交换机上划定 VLAN 和计算机的 IP 地址。如表 2-3-3 所示。

表 2-3-3　　　　　　　　交换机 VLAN 和计算机的 IP 地址

设备名称	VLAN 或接口	IP	子网掩码
Switch1	VLAN10（f0/1-10）	192.168.10.62	255.255.255.192
	VLAN20（f0/11-20）	192.168.10.126	255.255.255.192
PC1	网卡	192.168.10.1	255.255.255.192
PC2	网卡	192.168.10.2	255.255.255.192
PC3	网卡	192.168.10.65	255.255.255.192
PC4	网卡	192.168.10.66	255.255.255.192

步骤 2　配置交换机，创建 VLAN。

Switch>
Switch>enable　　　　　//进入特权用户模式
Switch#configure　terminal　　//进入全局配置模式
Switch(config)#hostname　Switch1　//配置交换机的名称为 Switch1
Switch1 (config)#vlan　10　　//创建 VLAN10
Switch1 (config-vlan)#name　xinxibu　//命名 VLAN10 为信息部的拼音形式
Switch1 (config-vlan)#exit　//返回上一级
Switch1(config)#interface　vlan　10　//进入 vlan10 接口
Switch1(config-if)#ip　address　192.168.10.62　255.255.255.192 //配置 vlan 10 的 IP 地址
Switch1(config-if)#no　shutdown　//开启接口
Switch1(config-if)#exit　//返回上一级
Switch1 (config)#vlan　20　　//创建 VLAN20
Switch1 (config-vlan)#name shefubu　//命名 VLAN20 为社服部的拼音形式
Switch1 (config-vlan)#exit

Switch1(config)#interface   vlan   20   //进入 vlan20 接口

Switch1(config-if)#ip   address   192.168.10.126   255.255.255.192 //配置 vlan20 的 IP 地址

Switch1(config-if)#no   shutdown   //开启接口

Switch1(config-if)#exit   //返回上一级

Switch1 (config)#interface   range   f0/1-10   //选中接口范围 f0/1-10

Switch1 (config-if-range)#switchport   access   vlan   10 //将该范围的接口划入 vlan10

Switch1 (config-if-range)#exit

Switch1 (config)#interface   range   f0/11-20   //选中接口范围 f0/1-10

Switch1 (config-if-range)#switchport   access   vlan   20   //将该范围的接口划入 vlan20

Switch1 (config-if-range)#exit

Switch1 (config)#

步骤 3 使用 PC1   ping   PC2，测试同一 VLAN 计算机的连通性。

PC>ipconfig   //查看本机 IP 信息

FastEthernet0 Connection:(default port)

IP Address.....................: 192.168.10.1

Subnet Mask....................: 255.255.255.192

Default Gateway................: 0.0.0.0

PC>ping   192.168.10.2   //测试 PC1 和 PC2 的连通性

Pinging 192.168.10.2 with 32 bytes of data:

Reply from 192.168.10.2: bytes=32 time=188ms TTL=128

Reply from 192.168.10.2: bytes=32 time=93ms TTL=128

Reply from 192.168.10.2: bytes=32 time=94ms TTL=128

Reply from 192.168.10.2: bytes=32 time=94ms TTL=128

Ping statistics for 192.168.10.2:

　　Packets: Sent = 4, Received = 4, Lost = 0 (0% loss),

Approximate round trip times in milli-seconds:

　　Minimum = 93ms, Maximum = 188ms, Average = 117ms

以上输出结果显示，同一 VLAN 的计算机可以互相 PING 通。

步骤 4 使用 PC1   ping   PC3，测试不同 VLAN 计算机的连通性。

PC>ping   192.168.10.65

Pinging 192.168.10.65 with 32 bytes of data:

Request timed out.

Request timed out.

Request timed out.

Request timed out.

Ping statistics for 192.168.10.65:

Packets: Sent = 4, Received = 0, Lost = 4 (100% loss)

我们发现，不同 VLAN 的计算机不能相互通信。

任务 2.4　跨交换机实现同一 VLAN 的主机通信

【实训目的】

掌握跨交换机同一 VLAN 中的计算机通信技术

【实训任务】

1. 根据拓扑图连接交换机和计算机，组建局域网。
2. 配置交换机，创建 VLAN。
3. 配置交换机的端口为 TRUNK 模式。
4. 测试跨交换机同一 VLAN 中计算机的通信

【预备知识】

交换机端口模式有 access 和 trunk 两种类型。Access 模式的端口只能属于一个 vlan，是交换机端口的默认模式。在 access 端口的链路上传输标准的以太网帧，即 IEEE802.3 数据帧，不附加任何标记，如图 2-4-1 所示。

| DA | SA | Type/Length | Data Frame | FCS |

图 2-4-1　标准的 802-3 数据帧结构

而 trunk 模式的端口同时属于多个 vlan，在两台交换机的 trunk 端口之间的链路上传输的是封装 802.1q 的数据帧（见图 2-4-2），在发送端打上标记，接收端再拆去标记。

| DA | SA | Tag | Type/Length | Data Frame | FCS |

图 2-4-2　封装 802.1q 后的数据帧结构

【实训拓扑】

实训拓扑图如图 2-4-3 所示。

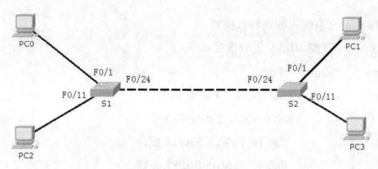

图 2-4-3 网络拓扑结构图

【实训设备】

二层交换机 2 台、计算机 2 台。

【实训步骤】

步骤 1 搭建拓扑，配置四台计算机的 IP 地址和子网掩码，如表 2-4 所示。

表 2-4

计算机	IP 地址	子网掩码	交换机	接口	VLAN
PC0	192.168.10.1	255.255.255.0	S1	F0/1	VLAN10
PC1	192.168.10.2	255.255.255.0	S2	F0/1	VLAN10
PC2	192.168.20.1	255.255.255.0	S1	F0/11	VLAN20
PC3	192.168.20.2	255.255.255.0	S2	F0/11	VLAN20

步骤 2 配置左侧交换机。

Switch>enable   //由用户模式进入特权模式

Switch#configure terminal //进入全局配置模式

Switch(config)#hostname s1 //更改交换机名称为 s1

s1(config)#vlan 10   //创建 VLAN10

s1(config-vlan)#exit   //返回到上一级的全局配置模式

s1(config)#vlan 20 //创建 VLAN20

s1(config-vlan)#exit   //返回到上一级的全局配置模式

  s1(config)#interface f0/1   //进入接口配置模式

s1(config-if)#switchport access vlan 10   //将 F0/1 划入 VLAN10

s1(config-if)#exit //返回到上一级模式

s1(config)#interface   f0/11

s1(config-if)#switchport access vlan 20   //将 F0/11 划入 VLAN20

s1(config-if)#end //直接返回到特权模式

s1#show vlan   //观察 VLAN 划分情况：

VLAN	Name	Status	Ports
1	default	active	Fa0/2, Fa0/3, Fa0/4, a0/5,
			Fa0/6, Fa0/7, Fa0/8, Fa0/9,
			Fa0/10, Fa0/12, Fa0/13, 0/14
			Fa0/15, Fa0/16, Fa0/17, 0/18,
			Fa0/19, Fa0/20, Fa0/21, Fa0/22,
			Fa0/23, Fa0/24
10	VLAN0010	active	Fa0/1
20	VLAN0020	active	Fa0/11

步骤 3 配置右侧交换机。

Switch>enable

Switch#configure terminal

Switch(config)#hostname s2

s2(config)#vlan 10

s2(config-vlan)#exit

s2(config)#vlan 20

s2(config-vlan)#exit

s2(config)#interface f0/1

s2(config-if)#switch access vlan 10

s2(config-if)#exit

s2(config)#interface f0/11

s2(config-if)#switchport access vlan 20

s2(config-if)#end

s2#show vlan

VLAN	Name	Status	Ports
1	default	active	Fa0/2, Fa0/3, Fa0/4, Fa0/5
			Fa0/6, Fa0/7, Fa0/8, Fa0/9
			Fa0/10, Fa0/12, Fa0/13, Fa0/14
			Fa0/15, Fa0/16, Fa0/17, Fa0/18
			Fa0/19, Fa0/20, Fa0/21, Fa0/22
			Fa0/23
10	VLAN0010	active	Fa0/1
20	VLAN0020	active	Fa0/11

步骤 4 测试连通性。

在命令提示符下测试 pc0 能否与另一交换机上同 vlan 的计算机 pc1 通信：

PC>ipconfig

IP Address----------------------: 192.168.10.1

Subnetwork Mask---------------------: 255.255.255.0

Default Gateway-----------------:0.0.0.0

PC>ping　192.168.10.2

Pinging 192.168.10.2 with 32 bytes of data:

Request timed out.

Request timed out.

Request timed out.

Request timed out.

Ping statistics for 192.168.10.2:

Packets: Sent = 4, Received = 0, Lost = 4 (100% loss)

结果显示：Pc0 与 pc1 不能正常通信。

步骤 5　配置两交换机之间的连接为 TRUNK 干道。

（1）配置左侧交换机。

　s1>enable

s1#configure　terminal

s1(config)#interface　f0/24

s1(config-if)#switch　mode　trunk　　//配置端口为 trunk 模式

s1(config-if)#end

（2）配置右侧交换机。

s2>

s2>enable

s2#configure terminal

s2(config)#interface　f0/24

s2(config-if)#switch　mode　trunk　　//配置端口为 trunk 模式

s2(config-if)#end

s2#

步骤 6　测试连通性。

在命令提示符下再次测试 pc0 能否与另一交换机上同 vlan 的计算机 pc1 通信：

PC>ping　192.168.10.2

Pinging 192.168.10.2 with 32 bytes of data:

Reply from 192.168.10.2: bytes=32 time=188ms TTL=128

Reply from 192.168.10.2: bytes=32 time=93ms TTL=128

Reply from 192.168.10.2: bytes=32 time=94ms TTL=128

Reply from 192.168.10.2: bytes=32 time=94ms TTL=128
Ping statistics for 192.168.10.2:

　　Packets: Sent = 4, Received = 4, Lost = 0 (0% loss),
Approximate round trip times in milli-seconds:

　　Minimum = 93ms, Maximum = 188ms, Average = 117ms

由此发现，配置 trunk 干道后，pc0 可以与 pc1 正常通信。

想一想：pc2 与 pc3 能通信吗？Pc0 能否与 pc3 正常通信呢？并试着分析其原因。

任务 2.5　交换机端口安全

【实训目的】

掌握交换机端口安全配置。

【实训任务】

1. 学会配置交换机端口的安全。
2. 配置交换机端口安全地址的绑定。
3. 限制交换机端口的最大连接数。

【预备知识】

目前的交换机大多都有端口安全功能，利用端口安全特性，可以实现网络接入安全。交换机端口安全机制是工作在交换机二层端口上的一个安全特性，它的功能是根据 MAC 地址来做对网络流量的控制和管理。比如 MAC 地址与具体的端口绑定，限制具体端口通过的 MAC 地址的数量；或者在具体的端口不允许某些 MAC 地址的帧流量通过。

当一个端口开启了端口安全特性以后，交换机将检查从此端口接收到的帧的源 MAC 地址，并检查在此端口上配置的最大安全地址数。若此帧的源 MAC 地址存在于安全地址表中，则直接转发帧；若安全地址数没有超过最大安全地址数量，并且此帧的源 MAC 地址不在安全地址表中，则交换机学习此 MAC 地址，将其添加到安全地址表中，进行后续转发。

交换机端口安全违例处理模式有以下三种。

Protect:当安全地址数目超过最大数目后，安全端口会丢弃来自非安全地址的数据帧。

Restrict:当违例发生时，安全端口会丢弃来自非安全地址的数据帧，并发送一个 trap 通知。

Shutdown:违例发生时，安全端口会丢弃来自非安全地址的数据帧，关闭端口并发送一个 trap 通知。

【实训拓扑】

网络拓扑结构图如图 2-5 所示。

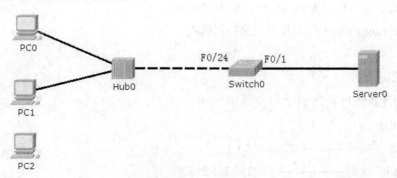

图 2-5 网络拓扑结构图

【实训设备】

交换机 1 台、集线器 1 台、计算机 4 台。

【实训步骤】

步骤 1 搭建网络，配置四台计算机的 IP 地址和子网掩码，如表 2-5 所示。

表 2-5

计算机	IP 地址	子网掩码	设备名称	接口
PC0	192.168.1.1	255.255.255.0	Hub0	Port 1
PC1	192.168.1.2	255.255.255.0	Hub0	Port 2
PC2	192.168.1.3	255.255.255.0	Hub0	
Server0	192.168.1.254	255.255.255.0	Switch0	F0/1

步骤 2 配置交换机端口安全。

Switch>

Switch>enable

Switch#

Switch#configure   terminal

Switch(config)#hostname   Switch0

Switch0(config)#interface   f0/24   //进入接口配置模式

Switch0(config-if)#switchport   mode   access   //设置交换机端口模式为 access

Switch0(config-if)#switchport   port-security   //开启端口安全

Switch0(config-if)#switchport port-security maximum 2
//设置接口上安全地址的最大数为 2，默认值为 128
Switch0(config-if)#switchport port-security violation shutdown
//设置处理违例的方式为 shutdown，默认为 protect
Switch0(config-if)#end　　//直接返回到特权模式
Switch0#

步骤 3　验证测试。

首先使用计算机 PC0 去 ping 服务器 Server0

PC>ipconfig

IP Address----------------------: 192.168.1.1

Subnetwork Mask----------------------: 255.255.255.0

Default Gateway------------------:0.0.0.0

PC>ping　192.168.1.254

Pinging 192.168.1.254 with 32 bytes of data:

Reply from 192.168.1.254: bytes=32 time=156ms TTL=128

Reply from 192.168.1.254: bytes=32 time=78ms TTL=128

Reply from 192.168.1.254: bytes=32 time=78ms TTL=128

Reply from 192.168.1.254: bytes=32 time=78ms TTL=128

Ping statistics for 192.168.1.254:

　　Packets: Sent = 4, Received = 4, Lost = 0 (0% loss),

Approximate round trip times in milli-seconds:

　　Minimum = 78ms, Maximum = 156ms, Average = 97ms

结果证实，计算机 Pc0 能够 ping 通服务器 Server0。

想一想：计算机 PC1 能否 ping 通服务器 Server0?

为了得到答案，我们进行验证：

PC>ipconfig

IP Address----------------------: 192.168.1.2

Subnetwork Mask----------------------: 255.255.255.0

Default Gateway------------------:0.0.0.0

PC>ping 192.168.1.254

Pinging 192.168.1.254 with 32 bytes of data:

Reply from 192.168.1.254: bytes=32 time=126ms TTL=128

Reply from 192.168.1.254: bytes=32 time=64ms TTL=128

Reply from 192.168.1.254: bytes=32 time=64ms TTL=128

Reply from 192.168.1.254: bytes=32 time=63ms TTL=128

Ping statistics for 192.168.1.254:

Packets: Sent = 4, Received = 4, Lost = 0 (0% loss),

Approximate round trip times in milli-seconds:

Minimum = 63ms, Maximum = 126ms, Average = 79ms

计算机 PC1 也能够 ping 通服务器 Server0。

此时，我们连接 PC2 到集线器的任一端口，测试其与服务器 Server0 的连通性。

PC>ipconfig

IP Address----------------------: 192.168.1.3

Subnetwork Mask---------------------: 255.255.255.0

Default Gateway-----------------:0.0.0.0

PC>ping 192.168.1.254

Pinging 192.168.1.254 with 32 bytes of data:

Request timed out.

Request timed out.

Request timed out.

Request timed out.

Ping statistics for 192.168.1.254:

　　Packets: Sent = 4, Received = 0, Lost = 4 (100% loss),

此时，由于连接计算机数超过了交换机接口 f0/24 的最大安全地址个数，该接口关闭。导致计算机 PC2 无法和服务器 server0 通信。

当端口由于违规操作而进入"err-disabled"状态后，必须在全局模式下使用如下命令手工将其恢复为 UP 状态：

Switch0(conifg)# errdisable　recovery

任务 2.6　配置快速生成树

【实训目的】

掌握配置快速生成树的方法。

【实训任务】

1. 根据拓扑连接网络设备，构建局域网。
2. 开启生成树。
3. 验证生成树。
4. 改变生成树的根桥。

【预备知识】

生成树协议 STP(Spanning Tree Protocol)的主要功能有两个：一是利用生成树算法，在以太网络中，创建一个以某台交换机的某个端口为根的生成树，避免环路；二是在以太网络拓扑发生变化时，通过生成树协议达到收敛保护的目的。

STP 中定义了根交换机、根端口、指定端口和路径开销等概念，意义在于构造一棵"树"的方法在达到冗余链路的同时，还实现了链路备份和路径最优化。用于构造"树"的算法被称为生成树算法。

交换机之间的冗余链路提高了网络的可靠性和稳定性。但也导致了广播风暴、多帧复制和地址表的不稳定等诸多问题。生成树通过软件协议判断网络中存在环路的地方，并暂时阻断冗余链路来实现。当主链路出现故障时，临时阻塞的链路马上恢复工作。

为了形成一个没有环路的拓扑，网络中的交换机要进行以下几个步骤：

①选举根桥；
②选举根口；
③选举指定口。

以上步骤中，选举顺序如下：

①最小的根桥 ID；
②最小的根路径代价；
③最小的发送者桥 ID；
④最小的发送者端口 ID。

【实训拓扑】

网络拓扑结构图如图 2-6 所示。

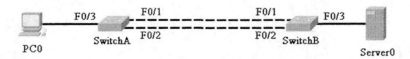

图 2-6 网络拓扑图

【实训设备】

二层交换机 2 台；计算机 2 台。

【实训步骤】

步骤 1 按时拓扑图构建网络，并配置计算机的 IP 地址，如表 2-6 所示。

表2-6

设备名称	接口	IP	子网掩码
PC0	网卡	192.168.1.1	255.255.255.0
Server0	网卡	192.168.1.2	255.255.255.0

步骤2 配置交换机SwitchA。

Switch>enable

Switch#configure terminal

Switch(config)#hostname SwitchA

SwitchA(config)#vlan 10

SwitchA(config-vlan)#name xinxibu

SwitchA(config-vlan)#exit

SwitchA(config)#interface f0/3

SwitchA(config-if)#switchport access vlan 10

SwitchA(config-if)#exit

SwitchA(config)#interface range f0/1-2

SwitchA(config-if-range)#switchport mode trunk

SwitchA(config-if-range)#end

步骤3 配置交换机SwitchB。

Switch>enable

Switch#configure terminal

Switch(config)#hostname SwitchB

SwitchB(config)#vlan 10

SwitchB(config-vlan)#name xinxibu

SwitchB(config-vlan)#exit

SwitchB(config)#interface f0/3

SwitchB(config-if)#switchport access vlan 10

SwitchB(config-if)#exit

SwitchB(config)#interface range f 0/1-2

SwitchB(config-if-range)#switchport mode trunk

SwitchB(config-if-range)#end

步骤4 配置快速生成树协议。

SwitchA#configure terminal

SwitchA(config)#spanning-tree   //开启生成树协议

SwitchA(config)#spanning-tree mode rstp   //指定生成树协议类型为RSTP

SwitchB#configure terminal

SwitchB(config)#spanning-tree   //开启生成树协议
SwitchB(config)#spanning-tree mode rstp   //指定生成树协议类型为 RSTP
验证快速生成树协议是否开启：
SwitchA#show spanning-tree   //查看生成树配置信息
StpVersion : RSTP   //生成树协议的版本为 RSTP
SysStpStatus : Enabled   //生成树协议的运行状态为 Enabled，已经开启；disable 为关闭状态
BaseNumPorts : 24
MaxAge : 20
HelloTime : 2
ForwardDelay : 15
BridgeMaxAge : 20
BridgeHelloTime : 2
BridgeForwardDelay : 15
MaxHops : 20
TxHoldCount : 3
PathCostMethod : Long
BPDUGuard : Disabled
BPDUFilter : Disabled
BridgeAddr : 00e1-f8bc-8824
Priority : 32768   //查看交换机的优先级，默认为 32768
TimeSinceTopologyChange : 0d:0h:1m:53s
TopologyChanges : 0
DesignatedRoot : 800000E1F8BC8824
RootCost : 0   //交换机到达根交换机的开销，0 表示此交换机为根
RootPort : 0   //查看交换机上的根端口，0 表示此交换机为根
SwitchB#show spanning-tree   //查看 SwitchB 生成树的配置信息
StpVersion : RSTP   //生成树协议的版本是 RSTP
SysStpStatus : Enabled   //生成树协议的运行状态为 Enabled，已经开启；disable 为关闭状态
BaseNumPorts : 24
MaxAge : 20
HelloTime : 2
ForwardDelay : 15
BridgeMaxAge : 20
BridgeHelloTime : 2
BridgeForwardDelay : 15
MaxHops : 20

TxHoldCount : 3
PathCostMethod : Long
BPDUGuard : Disabled
BPDUFilter : Disabled
BridgeAddr : 00e2-f8bf-faaa
Priority : 32768   //查看交换机的优先级，默认为 32768
TimeSinceTopologyChange : 0d:0h:2m:34s
TopologyChanges : 0
DesignatedRoot : 800000E1F8BC8824
RootCost : 200000   //交换机到达根交换机的开销为 200000
RootPort : Fa0/1   //查看交换机上的根端口，根端口为 Fa0/1

通过查看两台交换机的生成树信息发现，SwitchA 为根交换机，SwitchB 的 Fa0/1 为根端口。

步骤 5　设置交换机的优先级，指定 SwitchB 为根交换机。

SwitchB(config)#spanning-tree priority 4096   //设置交换机优先级为 4096
SwitchB(config)#end

验证测试：验证交换机 SwitchB 的优先级

SwitchB#show spanning-tree
StpVersion : RSTP
SysStpStatus : Enabled
BaseNumPorts : 24
MaxAge : 20
HelloTime : 2
ForwardDelay : 15
BridgeMaxAge : 20
BridgeHelloTime : 2
BridgeForwardDelay : 15
MaxHops : 20
TxHoldCount : 3
PathCostMethod : Long
BPDUGuard : Disabled
BPDUFilter : Disabled
BridgeAddr : 00e2-f8bf-faaa
Priority : 4096   //交换机的优先级为 4096
TimeSinceTopologyChange : 0d:0h:17m:3s
TopologyChanges : 0
DesignatedRoot : 100000E2F8BFFAAA

RootCost : 0

RootPort : 0

SwitchA#show spanning-tree   //查看 SwitchA 生成树的配置信息

StpVersion : RSTP   //生成树协议的版本为 RSTP

SysStpStatus : Enabled   //生成树协议已经开启，Disabled 为关闭状态

BaseNumPorts : 24

MaxAge : 20

HelloTime : 2

ForwardDelay : 15

BridgeMaxAge : 20

BridgeHelloTime : 2

BridgeForwardDelay : 15

MaxHops : 20

TxHoldCount : 3

PathCostMethod : Long

BPDUGuard : Disabled

BPDUFilter : Disabled

BridgeAddr : 00e1-f8bc-8824

Priority : 32768   //交换机的优先级为 32768

TimeSinceTopologyChange : 0d:0h:17m:19s

TopologyChanges : 0

DesignatedRoot : 100000 E2F8BFFAAA

RootCost : 200000   //交换机到达根交换机的开销是 200000

RootPort : Fa0/1   //交换机上的根端口是 Fa0/1

验证测试：验证交换机 SwitchA 的端口 1 和端口 2 的状态。

SwitchA#show spanning-tree interface FastEthernet 0/1
                //查看 SwitchA 端口 FastEthernet 0/1 的状态

PortAdminPortfast : Disabled

PortOperPortfast : Disabled

PortAdminLinkType : auto

PortOperLinkType : pointerface-to-pointerface

PortBPDUGuard: Disabled

PortBPDUFilter: Disabled

PortState : forwarding
        //SwitchA 的端口 FastEthernet 0/1 处于转发状态

PortPriority : 128

PortDesignatedRoot : 100000 E2F8BFFAAA

PortDesignatedCost : 0

PortDesignatedBridge : 100000 E2F8BFFAAA

PortDesignatedPort : 8001

PortForwardTransitions : 2

PortAdminPathCost : 0

PortOperPathCost : 200000

PortRole : rootPort   //端口角色为根端口

SwitchA#show spanning-tree interface FastEthernet 0/2

  //显示 SwitchA 端口 FastEthernet 0/2 的状态

PortAdminPortfast : Disabled

PortOperPortfast : Disabled

PortAdminLinkType : auto

PortOperLinkType : pointerface-to-pointerface

PortBPDUGuard: Disabled

PortBPDUFilter: Disabled

PortState : discarding

  //SwitchA 的端口 FastEthernet 0/2 处于丢弃状态

PortPriority : 128

PortDesignatedRoot : 100000 E2F8BFFAAA

PortDesignatedCost : 0

PortDesignatedBridge : 100000 E2F8BFFAAA

PortDesignatedPort : 8002

PortForwardTransitions : 1

PortAdminPathCost : 0

PortOperPathCost : 200000

PortRole : alternatePort   //SwitchA 的 F0/2 端口为根端口的替换端口

想一想：我们使用 PC0 不断地 ping 计算机 Server0，期间拔掉正常通信的网线，还能不能 ping 通？请自己验证测试，并查看丢包情况。

项目 3
配置路由器

路由器（Router）是连接因特网中各局域网、广域网的设备，它会根据信道的情况自动选择和设定路由，以最佳路径，按前后顺序发送信息。路由器是互联网络的"桥梁"。

路由和交换之间的主要区别就是交换发生在 OSI 参考模型第二层（数据链路层），而路由发生在第三层，即网络层。这一区别决定了路由和交换在移动信息的过程中需要使用不同的控制信息，所以两者实现各自功能的方式是不同的。

通过本项目的实训，我们可以理解路由的概念；学习简单配置路由器和设置路由器密码；掌握配置单臂路由的方法，实现各个 VLAN 间互相通信；了解静态路由和默认路由的作用及应用场合，掌握配置静态路由、默认路由和浮动路由的方法。

任务 3.1　简单配置路由器

【实训目的】

通过对路由器的简单配置，熟悉配置模式的切换，掌握路由器的基本使用。

【实训任务】

1. 掌握路由器各种配置模式的区别及模式之间的切换。
2. 熟练掌握路由器配置的基本命令。

【预备知识】

一、什么是路由器

路由器（Router）是一种计算机网络互连设备，工作在 OSI 模型的第三层——即网络层，用于连接因特网中各局域网、广域网。路由器能根据信道的情况自动选择和设定路由，以最佳路径将数据包通过一个个网络传送至目的地。路由器通常连接两个以上的网络。路由器是整个网络与外界的通信出口，也是联系内部子网的桥梁。

路由器缓存中存储着由网管员手工设置的或是路由协议自动创建的路由表。通过查看其路由表，路由器为接收到的每个数据包寻找一条最佳传输路径，并将该数据有效地传送到目的站点。路由器使数据分组到达其目的地的下一跳，而不是到达目的地的完整路径。在庞大的网络中可能存在很多的路径，路由器根据路由选择协议(RIP、OSPF 等)提供的功能，自动学习并记忆网络运行情况，在需要时自动计算数据传输的最佳路径，避开拥塞或暂时禁用链路的路径。

二、路由器的主要功能

路由器的主要功能就是"路由"的选择，通俗地讲就是"向导"作用，主要用来为数据包转发指明一个方向。路由器的"路由"功能可以细分为以下多个方面。

（1）在网际间接收节点发来的数据包，然后根据数据包中的源地址和目的地址，对照自己缓存中的路由表，把数据包直接转发到目的节点，这是路由器最主要也是最基本的路由作用。

（2）为网际间通信选择最合理的路由，这个功能其实是路由功能的扩展功能。假如有多个网络通过各自的路由器连在一起，而其中一个网络中的用户要向另一个网络的用户发出访问请求的话，路由器就会分析发出请求的源地址和接收请求的目的节点地址中的网络 ID 号，找出一条最经济、最快捷的通信路径。就像我们平时到了一个生疏的地点，不知道到目的地点的最佳走法，这时我们就得找一个向导，这个向导就会告诉我们一条最佳的路径，因为他熟悉各条路径的走法,这里所讲的路由器就相当于这个"向导"。

(3)拆分和包装数据包,这个功能也是路由功能的附属功能。因为有时在数据包转发流程中,由于网络带宽等因素,数据包过大的话很轻易造成网络堵塞,这时路由器就要把大的数据包根据对方网络带宽的状况拆分成小的数据包,到了目的网络的路由器后,目的网络的路由器会再把拆分的数据包装成一个与原来大小一致的数据包,再根据源网络路由器的转发信息获取目的节点的 MAC 地址,发给本地网络的节点。

(4)不同协议网络之间的连接。目前多数中、高档的路由器往往具有多通信协议支持的功能,这样就可以起到连接两个不同通信协议网络的作用。如常用 Windows NT 操作平台所运用的通信协议主要是 TCP/IP 协议,但假如是 NetWare 系统,则所采用的通信协议主要是 IPX/SPX 协议,这就需要依靠支持这些协议的路由器来连接。

(5)目前许多路由器都具有防火墙功能(可配置独立 IP 地址的网管型路由器),它能够起到基本的防火墙作用,也就是它能够屏蔽内部网络的 IP 地址,自由设定 IP 地址、通信端口过滤,使网络更加安全。

三、路由器工作模式

路由器的工作模式有三种:用户模式、特权模式和配置模式。

用于工作模式切换的命令归纳为图 3-1-1 所示。

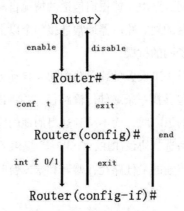

图 3-1-1 路由器工作模式切换命令示意图

【实训拓扑】

网络拓扑结构图如图 3-1-2 所示。

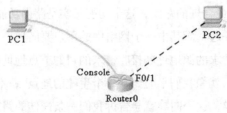

图 3-1-2 网络拓扑结构图

【实训设备】

PC 机 2 台、路由器 2 台、配置线 1 根。

【实训步骤】

步骤 1 构建本地配置环境（通过 Console 口配置）。

（1）将配置线一端连接到计算机的串口（即 COM 口），另一端连接到路由器的 Console 口。

（2）单击"开始"按钮，在"程序"菜单的"附件"选项中单击"超级终端"，弹出图 3-1-3 所示对话框。

图 3-1-3 超级终端界面

（3）在"名称"文本框中键入需要新建的超级终端连接项名称，然后单击"确定"按钮。弹出图 3-1-4 所示的"连接到"对话框。

图 3-1-4 "连接到"对话框

（4）在"连接时使用"下拉列表框中选择与路由器相连的计算机的串口，单击"确定"按钮，弹出图 3-1-5 所示的 COM1 属性对话框。

图 3-1-5　COM1 属性对话框

（5）在 COM1 属性中单击"还原为默认值"按钮，再单击"确定"按钮。

（6）按 Enter 键，出现路由器标识符：Router>

步骤 2　切换路由器配置模式。

Router>   //普通用户模式

Router >enable   //进入特权用户模式

Router #   //特权用户模式

Router #configure   terminal //进入全局配置模式

Router (config)#   //全局配置模式

Router (config)#interface   f0/1   //进入接口配置模式

Router (config-if)#   //接口配置模式

Router (config-if)#exit   //返回上一级模式

Router (config)#

Router (config)#exit   //返回上一级模式

Router #

Router #disable   //退出特权模式

Router >

Router >enable

Router # configure   terminal

Router (config)#interface   f0/1

Router (config-if)#end   //直接返回特权模式

Switch#

步骤 3　修改路由器的名字(Hostname)。

Router # configure   terminal

Router (config)#hostname　　Router0　　//修改路由器名字为"Router0"

步骤 4 设置路由器端口。

Router0 (config)#interface　　fastEthernet　　0/1

Router0 (config-if)#speed　　100　　//配置端口速率为 100Mbit/s

Router0 (config-if)# ip address　　192.168.1.1　　255.255.255.0　　//配置接口 IP 地址为 192.168.1.1/24

Router0 (config-if)#no　　shutdown　　//开启端口

Router0 (config-if)#exit

Router0 (config)#

步骤 5 管理路由器配置信息。

Router0#copy　　running-config　　startup-config //将当前运行的参数保存到启动文件中

Router0#write　　//将当前配置信息保存到 flash 中用于系统初始化

Router0#delete　　flash:config.text　　//永久删除 flash 中的配置文件

任务 3.2　配置路由器密码

【实训目的】

掌握路由器 Console 密码、Telnet 密码和 enable 密码的设置，加强路由器的安全管理。

【实训任务】

1. 配置路由器 Console 口和 vty 口。
2. 设置 enable 密码。
3. 配置本地认证。
4. 对所有的密码进行加密。

【预备知识】

一、设置路由器 Console 密码

路由器前面板有一个 Console 接口。我们用路由器附带的配置线一端接计算机的串口，另一端接到路由器的 console 接口，然后通过计算机自带的超级终端对路由器进行带外管理。

在全局配置模式下，设置控制台密码的命令如下：

Router(config)#line console 0

Router(config-line)#password 密码

Router(config-line)#login

二、设置路由器 Telnet 密码

设置路由器 Telnet 密码以后，我们就可以使用计算机自带的 telnet 工具远程登陆路由器，给网络设备的管理带来灵活性。

在全局配置模式下，设置 Telnet 密码的命令如下：

Router(config)#line vty 0 4

Router(config-line)#password 　密码

Router(config-line)#login

三、设置路由器 enable 密码

默认情况下，用户在路由器用户模式下输入 enable 命令，直接进入特权模式，就会拥有管理该路由器的所有权力，这样给安全管理带来了巨大的风险。设置路由器特权密码以后，当从用户模式进入特权模式时，需要用户输入正确的口令，验证通过才能进入特权模式，安全性大大增强。

特权密码又分加密密码和明文密码两种。加密密码优先级高于明文密码。如果同时设置了加密密码和明文密码，则明文密码失效。此外，加密密码无法正常显示，而明文密码可以在配置文件中看到。

在全局配置模式下，设置路由器特权密码的命令如下。

设置加密密码：Router(config)#enable　secret　　密码

设置明文密码：Router(config)#enable　password　　密码

四、本地认证 local

在 Console 口和 vty 口的 login 之后有 local 子命令，当启用 local 之后，必须设置本地用户的用户名和密码。在远程登录时输入用户名和密码进行认证，认证通过之后，才可以登录。

配置命令如下：

Router(config)#username　用户名　password　密码

Router(config)#line console 0

Router(config-line)#password 　密码

Router(config-line)#login local

Router(config-line)#exit

Router(config)#line vty 　0 4

Router(config-line)#password 　密码

Router(config-line)#login 　　local

五、加密所有明文密码

明文密码都会通过 show running-config 命令显示出来，这样会造成一定的安全隐患，我们可

以在全局配置模式下通过启用 service password-encryption 命令来对所有密码进行加密。命令如下：
Router(config)#service   password-encryption

【实训拓扑】

网络拓扑结构图如图 3-2-1 所示。

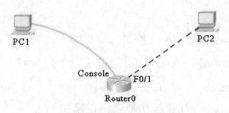

图 3-2-1　网络拓扑结构图

【实训设备】

路由器 1 台、计算机 2 台。

【实训步骤】

步骤 1　设置 Console 密码（使用 Packet Tracer）。

（1）我们在 PC1 上通过超级终端连入交换机，设置如图 3-2-2 所示。

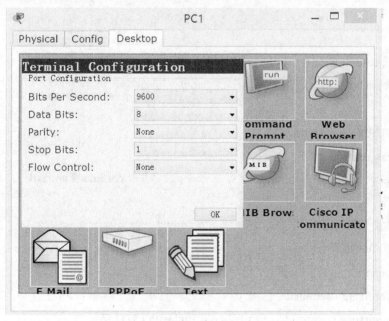

图 3-2-2　超级终端

单击"OK"按钮，进入路由器用户模式。

Router>

Router> enable   //由于尚未设置特权密码，可以直接进入特权模式

Router #

Router # configure terminal

Router(config)#hostname   Router0

（2）设置 console 密码。

Router0(config)#line console 0   //进入控制台口状态

Router0(config-line)#password   test1   //设置登录口令为 test1

Router0(config-line)#login   //允许登录

Router0(config-line)#end   //直接切换到特权模式

Router0#write   //保存配置信息

Building configuration---

[OK]

Router0#

（3）显示配置文件。

Router0#show   running-config

------

line   con   0

 password   test1   //可以看到设置的控制台密码 test1

 login

（4）测试 console 密码。

Router0#reload   //重启路由器

Press RETURN to get started!

User Access Verification

Password: //输入 console 密码 test1

Router0>

步骤 2 设置 Telnet 密码。

（1）配置 Telnet 密码。

Router0>enable

Router0#configure   terminal

Enter configuration commands, one per line.   End with CNTL/Z.

Router0(config)#line   vty   0   4   //进入虚拟终端配置模式

Router0(config-line)#password   test2   //设置登录口令 test2

Router0(config-line)#login   //允许登录
Router0(config-line)#exit   //返回上一层
（2）设置路由器接口 f0/1 的 IP 地址。
Router0(config)#interface   f0/1   //进入 f0/1 接口
Router0(config-if)#ip   address   192.168.1.1   255.255.255.0   //设置 IP 地址为 192.168.1.1
Router0(config-if)#no   shutdown   //开启接口
Router0(config-line)#exit   //返回上一层
（3）显示配置文件。
Router0#show   running-config
------
interface FastEthernet0/1
 ip address 192.168.1.1 255.255.255.0
------
line con 0
 password test1
 login
line vty 0 4
 password test2
 login

（4）测试 Telnet 密码。
首先配置 PC2 的 IP 地址为 192.168.1.10，子网掩码为 255.255.255.0。
然后再在 PC2 上运行 Telnet：
 PC>telnet   192.168.1.1
Trying 192.168.1.1 ---Open
User Access Verification
Password: //输入远程登录 Telnet 密码 test2
Router0>   //进入路由器用户模式
Router0>enable
% No password set.
Router0>   //由于尚未配置 enable 密码，所以远程用户无法进入特权模式
步骤 3 配置路由器 enable 密码。
Router0#configure   Terminal   //进入全局配置模式
Router0(config)#enable   secret test3   //设置加密密码为 test3
Router0(config)#enable   password   test4   //设置明文密码为 test4
Router0(config)#exit   //返回上一层模式

Router0#

Router0#disable   //退出特权模式

Router0>

（1）测试 enable 密码。

Router0>enable

Password: //输入明文密码 test4，无法进入特权模式

Password: //输入加密密码 test3，成功进入特权模式

Router0#

经过测试，明文密码和加密密码同时存在时，明文密码失效，表明加密密码优先级高于明文密码。

（2）显示配置文件。

Router0#show   running-config

------

enable secret 5 1mERr$Qvlp8arQQfzOuH-AEskzx0   //加密密码无法正常显示

enable password test4   //设置的明文密码 test4

步骤 4 配置本地认证。

（1）配置用户名和密码，并将 console、vty 口认证方式设为 local。

Router0(config)#username   qkzz   password qkzz

Router0(config)#line console   0

Router0(config-line)#login   local

Router0(config-line)#exit

Router0(config)#line   vty 0 4

Router0(config-line)#login   local

Router0(config-line)#end

Router0#

（2）显示配置。

Router0#show   running-config

------

hostname Router0

enable secret 5 1mERr$jmjjcNJfYnXCNd09Tn2dJ-

enable password test4

username qkzz password 0 qkzz

------

line con 0

 password test1   //登录密码 test1，在设置为本地认证后将失效

login local　　//登录认证方式为 local
line vty 0 4
　　password test2　　//登录密码 test2，在设置为本地认证后将失效
　　login local　　//登录认证方式为 local

（3）测试本地认证。

我们在 PC2 上运行 Telnet，测试结果如下：

PC>telnet 192.168.1.1

Trying 192.168.1.1 ---Open

User Access Verification

Username: qkzz　　//输入用户名 qkzz

Password:　　　　//尝试密码 test2,结果为失败。

% Login invalid

Username: qkzz　　//重新输入用户名 qkzz

Password:　　　　//输入密码 qkzz

Router0>　　　　//成功登录路由器

我们在特权模式下输入 reload，重启路由器，认证时也会要求输入用户名和密码，过程和上面相似。请同学们自己进行测试。

步骤 5　加密所有明文密码。

Router0(config)#service　password-encryption　　//加密明文密码，增强安全性

Router0#show　running-config

------

enable secret 5 1mERr$jmjjcNJfYnXCNd09Tn2dJ-

enable password 7 0835495D1D4D

------

username qkzz password 7 0830475413

------

line con 0

　password 7 0835495D1D48

　login local

line vty 0 4

　password 7 0835495D1D4B

　login local

以上输出结果表明，配置文件中的所有明文密码均被加密，无法正常显示。

任务 3.3 单臂路由

【实训目的】

1. 熟练配置交换机、路由器等设备。
2. 熟练配置单臂路由，实现不同 VLAN 的计算机互相通信。

【实训任务】

1. 按照拓扑图连接网络设备，构建局域网。
2. 在路由器上创建子接口，选择 VLAN 包封装格式，并激活路由选择协议。
3. 验证测试 VLAN 间的互通性。

【预备知识】

一、单臂路由

路由器一般用到最少 2 个物理接口，用来连接两个或多个不同网络，起到使网络互通的"桥梁"作用。当前，为了方便管理，在企业内部网络中通常划分出多个 VLAN。若 VLAN 之间的主机需要通信，而交换机又不支持三层交换，这时可以使用路由器的一个物理接口来解决多 VLAN 通信的问题，即平时所说的单臂路由。

二、连接交换机与路由器

用直通线连接交换机和路由器，交换机连接路由器的接口需设置为 trunk 模式。路由器端口默认为关闭，需要开启路由器接口，再在路由器子接口配置模式下封装 IEEE802.3 协议，并指明对应的 VLAN 号。

【实训拓扑】

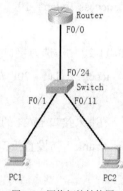

图 3-3 网络拓扑结构图

【实训设备】

二层交换机 1 台、路由器 1 台、计算机 2 台。

【实训步骤】

步骤 1 配置计算机 pc1 和 pc2 的 IP 地址以及网关地址，如表 3-3 所示。

表 3-3

设备名称	接口或 VLAN	IP 地址	子网掩码	网关
PC1	网卡	192.168.10.1	255.255.255.0	192.168.10.254
PC2	网卡	192.168.20.1	255.255.255.0	192.168.20.254
Switch	VLAN10（F0/1-10）			
	VLAN20（F0/11-20）			
Router	F 0/0.10	192.168.10.254	255.255.255.0	
	F 0/0.20	192.168.20.254	255.255.255.0	

步骤 2 配置交换机。

Switch>enable　　　　　　　　//进入特权模式

Switch#configure　terminal　　　//进入全局配置模式

Switch(config)#vlan　10　　　　//创建 vlan 10

Switch(config-vlan)#exit　　　　//返回上一级模式

Switch(config)#vlan　20　　　　//创建 vlan 20

Switch(config-vlan)#exit

Switch(config)#interface range f0/1-10   //选择接口范围从 f0/1 到 f0/10

Switch(config-if-range)#switchport access vlan 10   //将接口划入 vlan10

Switch(config-if-range)#exit

Switch(config)#interface range f0/11-20   //选择接口范围从 f0/11 到 f0/20

Switch(config-if-range)#switchport access vlan 20   //将接口划入 vlan20

Switch(config-if-range)#exit

Switch(config)#interface f0/24   //进入接口 f0/24 的配置模式

Switch(config-if)#switchport mode trunk   //设置接口为 trunk 模式

Switch(config-if)#exit

Switch(config)#

步骤 3 配置路由器。

Router>enable

Router#configure terminal

Router(config)#interface f0/0

Router(config-if)#no shutdown   //开启接口

Router(config-if)#exit

Router(config)#interface f0/0.10   //进入子接口 f0/0-10 的配置模式

Router(config-subif)#encapsulation dot1Q 10   //配置封装模式为 IEEE802.1q，对应 VLAN 号为 10

Router(config-subif)#ip address 192.168.10.254 255.255.255.0   //配置接口 IP 地址

Router(config-subif)#exit

Router(config)#interface f0/0.20   //进入子接口 f0/0-20 的配置模式

Router(config-subif)#encapsulation dot1Q 20   //配置封装模式为 IEEE802.1q，对应 VLAN 号为 20

Router(config-subif)#ip address 192.168.20.254 255.255.255.0   //配置接口 IP 地址

Router(config-subif)#exit

Router(config)#

步骤 4 查看路由表。

Router#show ip route

Codes: C - connected, S - static, I - IGRP, R - RIP, M - mobile, B - BGP

   D - EIGRP, EX - EIGRP external, O - OSPF, IA - OSPF inter area

   N1 - OSPF NSSA external type 1, N2 - OSPF NSSA external type 2

   E1 - OSPF external type 1, E2 - OSPF external type 2, E - EGP

   i - IS-IS, L1 - IS-IS level-1, L2 - IS-IS level-2, ia - IS-IS inter area

   * - candidate default, U - per-user static route, o - ODR

P - periodic downloaded static route

Gateway of last resort is not set

C   192.168.10.0/24 is directly connected, FastEthernet0/0.10
C   192.168.20.0/24 is directly connected, FastEthernet0/0.20

以上输出结果表明，路由器上有两条直连路由：192.168.10.0/24 直连到子接口 FastEthernet0/0.10；192.168.20.0/24 直连到 FastEthernet0/0.20。

步骤 5 测试 PC1 到 PC2 的连通性。

PC>ipconfig

IP Address----------------------: 192.168.10.1

Subnetwork Mask----------------------: 255.255.255.0

Default Gateway-----------------: 192.168.10.254

PC>ping 192.168.20.1

Pinging 192.168.20.1 with 32 bytes of data:

Request timed out.

Reply from 192.168.20.1: bytes=32 time=125ms TTL=127

Reply from 192.168.20.1: bytes=32 time=110ms TTL=127

Reply from 192.168.20.1: bytes=32 time=125ms TTL=127

Ping statistics for 192.168.20.1:

　　　Packets: Sent = 4, Received = 3, Lost = 1 (25% loss),

Approximate round trip times in milli-seconds:

　　　Minimum = 110ms, Maximum = 125ms, Average = 120ms

任务3.4 静态路由和默认路由

【实训目的】

1. 掌握静态路由和默认路由的配置命令。
2. 学会观察路由表。

【实训任务】

1. 根据拓扑图连接网络设备，构建局域网。
2. 配置静态路由。

3. 配置默认路由。
4. 测试网络互通性。

【预备知识】

一、静态路由

静态路由适用于较简单的网络环境，由管理员手工配置，不向外广播或组播，安全性、保密性高，同时节省了网络带宽。但是一旦某个网络节点出现故障，它不能随着网络的变化适时改变，必须由网络管理员手工修改相应的路由信息。

二、配置静态路由的过程

1.为路由器各个接口配置 IP 地址。
2.为每台路由器找出非直连的链路地址。
3.为每台路由器逐一写出到达各非直连网络地址的路由命令。

三、静态路由命令格式

ip　route　目的地址　　子网掩码　　下一跳地址 | 网络出口

四、默认路由

默认路由通常用于末端网络，即只有一个网络出口的网络。
默认路由命令格式：ip　route　0.0.0.0　0.0.0.0　下一跳 | 网络出口

【实训拓扑】

网络拓扑结构图如图 3-4 所示。

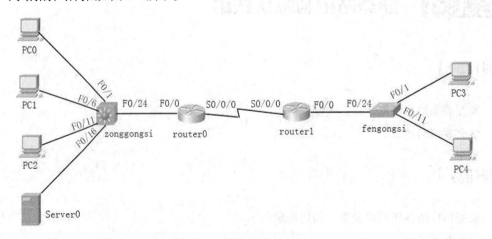

图 3-4　网络拓扑结构图

【实训设备】

路由器 2 台、三层交换机 1 台、二层交换机 1 台、计算机 6 台。

【实训步骤】

步骤 1 配置各网络设备的接口 IP 地址，参数见表 3-4 所示。

表 3-4

设备名称	接口	IP 地址	子网掩码	网关
Pc0	网卡	192.168.10.2	255.255.255.0	192.168.10.1
Pc1	网卡	192.168.20.2	255.255.255.0	192.168.20.1
Pc2	网卡	192.168.30.2	255.255.255.0	192.168.30.1
Pc3	网卡	172.16.50.2	255.255.255.0	172.16.50.1
Pc4	网卡	172.16.60.2	255.255.255.0	172.16.60.1
Server0	网卡	192.168.40.2	255.255.255.0	192.168.40.1
zonggongsi	Vlan10 (f0/1-5)	192.168.10.1	255.255.255.0	
	Vlan20 (f0/6-10)	192.168.20.1	255.255.255.0	
	Vlan30 (f0/11-15)	192.168.30.1	255.255.255.0	
	Vlan40 (f0/16-20)	192.168.40.1	255.255.255.0	
	F0/24	10.0.0.1	255.255.255.252	
Router0	F0/0	10.0.0.2	255.255.255.252	
	S0/0/0	10.0.0.5	255.255.255.252	
Router1	S0/0/0	10.0.0.6	255.255.255.252	
	F0/0-10	172.16.50.1	255.255.255.0	
	F0/0-20	172.16.60.1	255.255.255.0	
fengongsi	Vlan50 (f0/1-10)			
	Vlan60 (f0/11-20)			

(1) 配置三层交换机 zonggongsi。

Switch>

Switch>enable

Switch#configure Terminal

Switch(config)#hostname zonggongsi

zonggongsi (config)#vlan 10

zonggongsi (config-vlan)#exit

zonggongsi (config)#vlan 20

zonggongsi (config-vlan)#exit

zonggongsi (config)#vlan 30

zonggongsi (config-vlan)#exit

zonggongsi (config)#vlan 40

zonggongsi (config-vlan)#exit

zonggongsi (config)#interface range f0/1-5

zonggongsi (config-if-range)#switchport access vlan 10

zonggongsi (config-if-range)#exit

zonggongsi (config)#interface range f0/6-10

zonggongsi (config-if-range)#switchport access vlan 20

zonggongsi (config-if-range)#exit

zonggongsi (config)#interface range f0/11-15

zonggongsi (config-if-range)#switchport access vlan 30

zonggongsi (config-if-range)#exit

zonggongsi (config)#interface range f0/16-20

zonggongsi (config-if-range)#switchport access vlan 40

zonggongsi (config-if-range)#exit

zonggongsi(config)#interface vlan 10

zonggongsi(config-if)#ip address 192.168.10.1 255.255.255.0

zonggongsi(config-if)#exit

zonggongsi(config)#interface vlan 20

zonggongsi(config-if)#ip address 192.168.20.1 255.255.255.0

zonggongsi(config-if)#exit

zonggongsi(config)#interface vlan 30

zonggongsi(config-if)#ip address 192.168.30.1 255.255.255.0

zonggongsi(config-if)#exit

zonggongsi(config)#interface vlan 40

zonggongsi(config-if)#ip address 192.168.40.1 255.255.255.0

zonggongsi(config-if)#exit

zonggongsi(config)#interface   f0/24

zonggongsi(config-if)#no   switchport   //将接口转化为路由口

zonggongsi(config-if)#ip   address   10.0.0.1   255.255.255.252

zonggongsi(config-if)#exit

zonggongsi(config)#ip   routing   //开启三层交换机的路由功能

验证测试：

zonggongsi#show   ip   interface   brief   // 观察各接口 ip 配置信息

Interface	IP-Address	OK? Method Status	Protocol
FastEthernet0/24	10.0.0.1	YES manual up	up
Vlan10	192.168.10.1	YES manual up	up
Vlan20	192.168.20.1	YES manual up	up
Vlan30	192.168.30.1	YES manual up	up
Vlan40	192.168.40.1	YES manual up	up

（2）配置路由器 router0 的接口 IP。

Router>enable

Router#configure   terminal

Router(config)#hostname   router0

router0(config)#interface   f0/0

router0(config-if)#ip address   10.0.0.2   255.255.255.252

router0(config-if)#no   shutdown

router0(config-if)#exit

router0(config)#interface   s0/0/0

router0(config-if)#ip   address   10.0.0.5   255.255.255.252

router0(config-if)#clock   rate   128000

router0(config-if)#no   shutdown

router0(config-if)#exit

router0(config)#

（3）配置路由器 router1 的接口 IP。

Router>enable

Router#configure   terminal

Router(config)#hostname   router1

router1(config)#interface   s0/0/0

router1(config-if)#ip   address   10.0.0.6   255.255.255.252

router1(config-if)#no   shutdown

router1(config-if)#exit

router1(config)#interface f0/0

router1(config-if)#no shutdown

router1(config-if)#exit

router1(config)#interface f0/0.10 //进入子接口 f0/0.10 的配置模式

router1(config-subif)#encapsulation dot1q 50 //配置封装模式为 IEEE802.1q，对应 VLAN 号为 50

router1(config-subif)#ip address 172.16.50.1 255.255.255.0

router1(config-subif)#exit

router1(config)#interface f0/0.20 //进入子接口 f0/0.20 的配置模式

router1(config-subif)#encapsulation dot1q 60 //配置封装模式为 IEEE802.1q，对应 VLAN 号为 60

router1(config-subif)#ip address 172.16.60.1 255.255.255.0

router1(config-subif)#exit

（4）配置二层交换机 fengongsi。

Switch>enable

Switch#configure terminal

Switch(config)#hostname fengongsi

fengongsi(config)#vlan 50

fengongsi(config-vlan)#exit

fengongsi(config)#vlan 60

fengongsi(config-vlan)#exit

fengongsi(config)#interface range f0/1-10

fengongsi(config-if-range)#switchport access vlan 50

fengongsi(config-if-range)#exit

fengongsi(config)#interface range f0/11-20

fengongsi(config-if-range)#switchport access vlan 60

fengongsi(config-if-range)#exit

fengongsi(config)#interface f0/24

fengongsi(config-if)#switchport mode trunk

fengongsi(config-if)#end

fengongsi#

（5）配置各计算机的 IP 地址、子网掩码和网关。

步骤2 配置各网络设备的静态路由和默认路由。

（1）配置 zonggongsi 的默认路由。

zonggongsi(config)#ip route 0.0.0.0 0.0.0.0 10.0.0.2

（2）配置 router0 的静态路由。

```
router0(config)#ip    route    192.168.10.0    255.255.255.0    10.0.0.1
router0(config)#ip    route    192.168.20.0    255.255.255.0    10.0.0.1
router0(config)#ip    route    192.168.30.0    255.255.255.0    10.0.0.1
router0(config)#ip    route    192.168.40.0    255.255.255.0    10.0.0.1
router0(config)#ip    route    172.16.50.0     255.255.255.0    10.0.0.6
router0(config)#ip    route    172.16.60.0     255.255.255.0    10.0.0.6
```

（3）配置 router1 的默认路由。

```
router1(config)#ip    route 0.0.0.0    0.0.0.0    10.0.0.5
```

请思考：为什么 zonggongsi 和 router1 配置默认路由？

步骤 3 观察各网络设备的路由表信息。

```
zonggongsi#show    ip    route
Codes: C - connected, S - static, I - IGRP, R - RIP, M - mobile, B - BGP
       D - EIGRP, EX - EIGRP external, O - OSPF, IA - OSPF inter area
       N1 - OSPF NSSA external type 1, N2 - OSPF NSSA external type 2
       E1 - OSPF external type 1, E2 - OSPF external type 2, E - EGP
       i - IS-IS, L1 - IS-IS level-1, L2 - IS-IS level-2, ia - IS-IS inter area
       * - candidate default, U - per-user static route, o - ODR
       P - periodic downloaded static route

Gateway of last resort is 10.0.0.2 to network 0.0.0.0

        10.0.0.0/30 is subnetted, 1 subnets
C          10.0.0.0 is directly connected, FastEthernet 0/24
C       192.168.10.0/24 is directly connected, Vlan10
C       192.168.20.0/24 is directly connected, Vlan20
C       192.168.30.0/24 is directly connected, Vlan30
C       192.168.40.0/24 is directly connected, Vlan40
S*      0.0.0.0/0 [1/0] via 10.0.0.2
```

以上输出结果表明，总公司交换机上有 5 条直连路由和 1 条默认路由。直连路由无须设置，由设备自动学习生成。

```
router0#show    ip    route
Codes: C - connected, S - static, I - IGRP, R - RIP, M - mobile, B - BGP
       D - EIGRP, EX - EIGRP external, O - OSPF, IA - OSPF inter area
       N1 - OSPF NSSA external type 1, N2 - OSPF NSSA external type 2
       E1 - OSPF external type 1, E2 - OSPF external type 2, E - EGP
       i - IS-IS, L1 - IS-IS level-1, L2 - IS-IS level-2, ia - IS-IS inter area
```

         * - candidate default, U - per-user static route, o - ODR
         P - periodic downloaded static route

Gateway of last resort is not set

         10.0.0.0/30 is subnetted, 2 subnets
C   10.0.0.0 is directly connected, FastEthernet0/0
C   10.0.0.4 is directly connected, Serial0/0/0
         172.16.0.0/24 is subnetted, 2 subnets
S   172.16.50.0 [1/0] via 10.0.0.6
S   172.16.60.0 [1/0] via 10.0.0.6
S   192.168.10.0/24 [1/0] via 10.0.0.1
S   192.168.20.0/24 [1/0] via 10.0.0.1
S   192.168.30.0/24 [1/0] via 10.0.0.1
S   192.168.40.0/24 [1/0] via 10.0.0.1

以上输出结果显示，router0 上有我们配置的 6 条静态路由和设备自动学习生成的 2 条直连路由。
router1#show ip route
Codes: C - connected, S - static, I - IGRP, R - RIP, M - mobile, B - BGP
       D - EIGRP, EX - EIGRP external, O - OSPF, IA - OSPF inter area
       N1 - OSPF NSSA external type 1, N2 - OSPF NSSA external type 2
       E1 - OSPF external type 1, E2 - OSPF external type 2, E - EGP
       i - IS-IS, L1 - IS-IS level-1, L2 - IS-IS level-2, ia - IS-IS inter area
       * - candidate default, U - per-user static route, o - ODR
       P - periodic downloaded static route

Gateway of last resort is 10.0.0.5 to network 0.0.0.0

         10.0.0.0/30 is subnetted, 1 subnets
C   10.0.0.4 is directly connected, Serial0/0/0
         172.16.0.0/24 is subnetted, 2 subnets
C   172.16.50.0 is directly connected, FastEthernet0/0-10
C   172.16.60.0 is directly connected, FastEthernet0/0-20
S*   0.0.0.0/0 [1/0] via 10.0.0.5

以上输出结果显示，router1 上有我们配置的 1 条默认路由和路由器自动学习生成的 3 条直连路由。

步骤 4 验证测试 pc0 和 pc4 的互通性。

PC>ping   172.16.60.2

Pinging 172.16.60.2 with 32 bytes of data:

Reply from 172.16.60.2: bytes=32 time=157ms TTL=125

Reply from 172.16.60.2: bytes=32 time=141ms TTL=125

Reply from 172.16.60.2: bytes=32 time=140ms TTL=125

Reply from 172.16.60.2: bytes=32 time=143ms TTL=125

Ping statistics for 172.16.60.2:

　　Packets: Sent = 4, Received = 4, Lost = 0 (0% loss),

Approximate round trip times in milli-seconds:

Minimum = 140ms, Maximum = 157ms, Average = 145ms

经过测试，总公司的 pc0 能够和分公司的 pc4 互相通信。

想一想：其他设备能互通吗？请自己验证测试，并分析数据包是如何选路并传送到目的设备上的。

任务 3.5　浮动静态路由

【实训目的】

配置浮动静态路由，构建可靠网络环境。

【实训任务】

1. 配置静态路由。
2. 配置浮动静态路由。
3. 验证测试浮动路由。

【预备知识】

一、常见路由类型的管理距离

路由类型	管理距离
直连路由	0
静态路由	1
RIP 路由	120
OSPF 路由	110

二、浮动路由

网络管理员可以手工修改管理距离来指定最佳路径，让另一条传输路径暂时隐藏，当设定的最佳路径断开时，隐藏的路径浮现出来，所以叫作浮动路由。

设定浮动静态路由的语法为

Router(config)#ip route 目的网络 子网掩码 下一跳地址 管理距离

【实训拓扑】

网络拓扑结构图如图 3-5 所示。

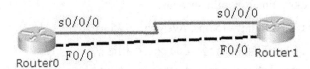

图 3-5 网络拓扑结构图

【实训设备】

路由器 2 台。

【实训步骤】

步骤 1 根据拓扑结构图，组建网络，并配置路由器接口地址，如表 3-5 所示。

表 3-5

设备	接口	IP 地址	子网掩码
Router0	Loopback0	172.16.1.1	255.255.255.0
	S0/0/0	12.1.1.1	255.255.255.252
	F0/0	13.1.1.1	255.255.255.252
Router1	Loopback0	192.168.1.1	255.255.255.252
	S0/0/0	12.1.1.2	255.255.255.252
	F0/0	13.1.1.2	255.255.255.252

（1）配置路由器 Router0 接口的 IP 地址，并开启接口。

Router>enable

Router#configrue terminal

Router(config)#hostname Router0

Router0(config)#interface s0/0/0

Router0(config-if)#ip   address   12.1.1.1 255.255.255.252

Router0(config-if)#clock   rate   128000

Router0(config-if)#no   shutdown

Router0(config-if)#exit

Router0(config)#interface   loopback 0

Router0(config-if)#ip   address   172.16.1.1   255.255.255.0

Router0(config-if)#exit

（2）配置路由器 Router1 接口的 IP 地址，并开启接口。

Router>enable

Router#configure   terminal

Router(config)#hostname   Router1

Router1(config)#interface   s0/0/0

Router1(config-if)#ip   address   12.1.1.2 255.255.255.252

Router1(config-if)#no   shutdown

Router1(config-if)#exit

Router1(config)#interface   loopback   0

Router1(config-if)#ip   address   192.168.1.1   255.255.255.0

Router(config-if)#exit

步骤 2 配置静态路由和浮动静态路由。请务必设置回指路由。

（1）配置路由器 Router0 的静态路由。

Router0(config)#ip   route   192.168.1.0 255.255.255.0   13.1.1.2

Router0(config)#ip   route   192.168.1.0   255.255.255.0   12.1.1.2   50   //设定管理距离为 50

（2）配置路由器 Router1 的静态路由。

Router1(config)#ip   route   172.16.1.0   255.255.255.0   13.1.1.1

Router1(config)#ip   route   172.16.1.0   255.255.255.0   12.1.1.1   50 //设定管理距离为 50

步骤 3 查看路由表。

（1）查看路由器 Router0 的路由表。

Router0#show   ip   route

Codes: C - connected, S - static, I - IGRP, R - RIP, M - mobile, B - BGP

       D - EIGRP, EX - EIGRP external, O - OSPF, IA - OSPF inter area

       N1 - OSPF NSSA external type 1, N2 - OSPF NSSA external type 2

       E1 - OSPF external type 1, E2 - OSPF external type 2, E - EGP

       i - IS-IS, L1 - IS-IS level-1, L2 - IS-IS level-2, ia - IS-IS inter area

       * - candidate default, U - per-user static route, o - ODR

       P - periodic downloaded static route

Gateway of last resort is not set

    12.0.0.0/30 is subnetted, 1 subnets
C   12.1.1.0 is directly connected, Serial0/0/0
    13.0.0.0/30 is subnetted, 1 subnets
C   13.1.1.0 is directly connected, FastEthernet0/0
    172.16.0.0/24 is subnetted, 2 subnets
C   172.16.1.0 is directly connected, Loopback0
C   172.16.2.0 is directly connected, Loopback1
S   192.168.1.0/24 [1/0] via 13.1.1.2

我们只发现了一条静态路由"S   192.168.1.0/24 [1/0] via 13.1.1.2"，下一跳地址是 13.1.1.2。而我们设定的下一跳地址为 12.1.1.2 的静态路由被隐藏了。

（2）查看路由器 Router1 的路由表。

Router1#show   ip   route
Codes: C - connected, S - static, I - IGRP, R - RIP, M - mobile, B - BGP
      D - EIGRP, EX - EIGRP external, O - OSPF, IA - OSPF inter area
      N1 - OSPF NSSA external type 1, N2 - OSPF NSSA external type 2
      E1 - OSPF external type 1, E2 - OSPF external type 2, E - EGP
      i - IS-IS, L1 - IS-IS level-1, L2 - IS-IS level-2, ia - IS-IS inter area
      * - candidate default, U - per-user static route, o - ODR
      P - periodic downloaded static route

Gateway of last resort is not set

    12.0.0.0/30 is subnetted, 1 subnets
C   12.1.1.0 is directly connected, Serial0/0/0
    13.0.0.0/30 is subnetted, 1 subnets
C   13.1.1.0 is directly connected, FastEthernet0/0
    172.16.0.0/24 is subnetted, 1 subnets
S   172.16.1.0 [1/0] via 13.1.1.1
C   192.168.1.0/24 is directly connected, Loopback0

同样，我们在 Router1 上只发现了一条静态路由"S   172.16.1.0 [1/0] via 13.1.1.1"，下一跳地址是 13.1.1.1。而我们设定的下一跳地址为 12.1.1.1 的静态路由被隐藏了。

步骤 4 测试连通性。

我们在路由器 Router0 上进行测试。

Router0#ping

Protocol [ip]:

Target IP address: 192.168.1.1

Repeat count [5]:

Datagram size [100]:

Timeout in seconds [2]:

Extended commands [n]: y

Source address or interface: 172.16.1.1

Type of service [0]:

Set DF bit in IP header? [no]:

Validate reply data? [no]:

Data pattern [0xABCD]:

Loose, Strict, Record, Timestamp, Verbose[none]:

Sweep range of sizes [n]:

Type escape sequence to abort.

Sending 5, 100-byte ICMP Echos to 192.168.1.1, timeout is 2 seconds:

Packet sent with a source address of 172.16.1.1

!!!!!

Success rate is 100 percent (5/5), round-trip min/avg/max = 31/31/32 ms

现在全网能够互联互通了。

想一想：假设我们将 Router0 的接口 f0/0 关闭，全网还能互通吗？

步骤 5　验证浮动静态路由。

（1）关闭 Router0 的接口 f0/0。

Router0(config)#interface   f0/0

Router0(config-if)#shutdown

（2）查看路由器 Router0 的路由表。

Router#show   ip   route

Codes: C - connected, S - static, I - IGRP, R - RIP, M - mobile, B - BGP

　　　　D - EIGRP, EX - EIGRP external, O - OSPF, IA - OSPF inter area

　　　　N1 - OSPF NSSA external type 1, N2 - OSPF NSSA external type 2

　　　　E1 - OSPF external type 1, E2 - OSPF external type 2, E - EGP

　　　　i - IS-IS, L1 - IS-IS level-1, L2 - IS-IS level-2, ia - IS-IS inter area

　　　　* - candidate default, U - per-user static route, o - ODR

　　　　P - periodic downloaded static route

Gateway of last resort is not set

```
        12.0.0.0/30 is subnetted, 1 subnets
C          12.1.1.0 is directly connected, Serial0/0/0
        172.16.0.0/24 is subnetted, 1 subnets
C          172.16.1.0 is directly connected, Loopback0
S       192.168.1.0/24 [50/0] via 12.1.1.2
```

经过观察，被隐藏的静态路由"S   192.168.1.0/24 [50/0] via 12.1.1.2"显示出来了。

（3）测试连通性。

在路由器 Router0 上进行测试。

Router0#ping

Protocol [ip]:

Target IP address: 192.168.1.1

Repeat count [5]:

Datagram size [100]:

Timeout in seconds [2]:

Extended commands [n]: y

Source address or interface: 172.16.1.1

Type of service [0]:

Set DF bit in IP header? [no]:

Validate reply data? [no]:

Data pattern [0xABCD]:

Loose, Strict, Record, Timestamp, Verbose[none]:

Sweep range of sizes [n]:

Type escape sequence to abort.

Sending 5, 100-byte ICMP Echos to 192.168.1.1, timeout is 2 seconds:

Packet sent with a source address of 172.16.1.1

!!!!!

Success rate is 100 percent (5/5), round-trip min/avg/max = 31/31/32 ms

在网络管理员指定的最佳路径断掉后，浮动路由起到了传输数据的重要作用。由此可见，通过巧妙设置浮动路由，增加一条冗余链路，网络的可靠性得到了更有效的保障。

项目 4 动态路由协议

动态路由是路由器根据网络系统的运行情况而自动计算调整的路由。相互连接的路由器之间互相交换彼此的信息，然后按照一定的算法计算出路由表，并且路由信息是在一定时间间隙里不断更新的，这样才能适应不断变化的网络，随时获得最优的寻路效果。

动态路由协议分为距离矢量路由协议和链路状态路由协议。RIP 协议是典型的距离矢量路由协议，OSPF 是典型的链路状态路由协议。

通过本项目的实训，我们可以了解动态路由和静态路由的特点、区别，掌握动态路由协议 RIP 和 OSPF 的配置方法。

任务 4.1 动态路由协议 RIPV2

【实训目的】

配置 RIP 协议，了解动态路由的原理，区分静态路由和动态路由。

【实训任务】

1. 根据拓扑连接网络设备，构建局域网。
2. 配置 RIP 协议，观察路由表。
3. 验证测试网络连通性。

【预备知识】

一、RIP 路由信息协议

RIP 是被广泛使用的距离向量路由协议，适用于小型网络环境。运行 RIP 功能的路由器默认每隔 30s 与它直连的网络邻居广播（RIPV1）或组播（RIPV2）路由更新。路由学习和更新将产生较大的流量，占用过多的带宽。

作为距离矢量路由协议，RIP 使用距离矢量来决定最佳路径。简单地说，就是使用从源网段到目的网段所经过的路由器的个数（跳数）来计算度量值，每经过一个路由器，跳数就增加 1。RIP 路由协议认为跳数越少，路径的路由就越佳，并将其加入路由表。如果到达目的网段有两条或多条跳数相等的路径，尽管带宽不同，RIP 也会认为它们是等价路径。RIP 默认支持 4 条等价路径，最大支持 6 条等价路径。

RIP 使用一些时钟来保证它所维持的路由的有效性和及时性。

1. 更新计时器（Update Timer）

平均每 30s 发送一个响应消息。为了防止更新同步，RIP 会以 15%的误差发送更新，即实际发送更新的周期范围是 25.5~30s。

2. 无效计时器（Invalid Timer）

当有一条新的路由被建立，无效计时器就会被设置为 180s，每当接收到这条路由的更新后，计时器又将重置为初始值。如果一条路由的更新在 180s 内还没有收到，则它将被标记为不可到达——该路由的度量被设置为 16。

3. 刷新计时器（Flush Timer）

刷新计时器的时间默认为 240s，可理解为若在某路由被标记为不可到达后的 60s 仍没有收到该路由的 RIP 消息，就将该路由从路由表中删除。

4. 抑制计时器（Holddown Timer）

如果一条路由更新的跳数大于路由表已记录的跳数，则该路由进入长达 180s 的抑制状态，以防止 RIP 中可能发生的路由环路。

RIP 路由协议配置简单，适合小型网络。对企业网络管理员来说，学习和使用 RIP 路由协议非常容易，降低了配置静态路由的复杂度。

二、RIP 协议的特征

（1）是距离向量路由协议。
（2）使用跳数作为度量值，最大跳数为 15 跳。
（3）默认路由更新周期为 30s，支持触发更新。
（4）管理距离为 120。

三、RIPV2 的配置命令

router (config)#router   rip   //启用路由协议 RIP
router (config-router)#version   2   //RIP 的版本为 2
router (config-router)#no   auto-summary   //取消自动汇总功能
router (config-router)#network   主类网络号

【实训拓扑】

网络拓扑结构图如图 4-1 所示。

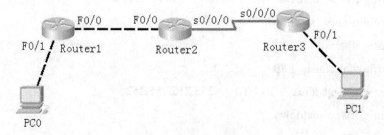

图 4-1 网络拓扑结构图

【实训设备】

路由器 3 台、计算机 2 台。

95

【实训步骤】

步骤 1 配置各网络设备接口 IP，参见表 4-1 所示。

表 4-1

设备名称	接口	IP 地址	子网掩码
Router1	F0/1	192.168.10.1	255.255.255.0
	F0/0	10.1.1.1	255.255.255.252
Router2	F0/0	10.1.1.2	255.255.255.252
	S0/0/0	10.1.1.5	255.255.255.252
Router3	S0/0/0	10.1.1.6	255.255.255.252
	F0/1	192.168.20.1	255.255.255.0
Pc0	网卡	192.168.10.2	255.255.255.0
Pc1	网卡	192.168.20.2	255.255.255.0

（1）配置 router1 的接口 IP。

Router>enable

Router#configure Terminal

Router(config)#hostname router1

router1(config)#interface f0/1

router1(config-if)#ip address 192.168.10.1 255.255.255.0

router1(config-if)#no shutdown

router1(config-if)#exit

router1(config)#interface f0/0

router1(config-if)#ip address 10.1.1.1 255.255.255.252

router1(config-if)#no shutdown

router1(config-if)#exit

router1(config)#

（2）配置 router2 的接口 IP。

Router>enable

Router#configure terminal

Router(config)#hostname router2

router2(config)#interface f0/0

router2(config-if)#ip address 10.1.1.2 255.255.255.252

router2(config-if)#no shutdown

项目 4　动态路由协议

router2(config-if)#exit

router2(config)#interface　s0/0/0

router2(config-if)#ip　address　10.1.1.5　255.255.255.252

router2(config-if)#clock　rate　128000　//DCE 端切记要配置时钟频率

router2(config-if)#no　shutdown

router2(config-if)#exit

router2(config)#

（3）配置 router3 的接口 IP。

Router>enable

Router#configure　terminal

Router(config)#hostname　router3

Router3(config)#interface　s0/0/0

Router3(config-if)#ip　address　10.1.1.6　255.255.255.252

Router3(config-if)#no　shutdown

Router3(config-if)#exit

Router3(config)#interface　f0/1

Router3(config-if)#ip　address　192.168.20.1　255.255.255.0

Router3(config-if)#no　shutdown

Router3(config-if)#exit

Router3(config)#

步骤 2　配置路由协议 RIP。

（1）配置路由器 router1，启用 RIP 协议。

router1(config)#router　rip　　　//启用路由协议 RIP

router1(config-router)#version　2　　//RIP 的版本为 2

router1(config-router)#no　auto-summary　　//取消自动汇总功能

router1(config-router)#network　192.168.10.0　//通告网络 192.168.10.0

router1(config-router)#network　10.0.0.0　//RIP 是有类路由协议，只通告主类网络即可。

router1(config-router)#end

router1#

（2）配置路由器 router2，启用 RIP 协议。

router2(config)#router　rip

router2(config-router)#version　2

router2(config-router)#no　auto-summary

router2(config-router)#network　10.0.0.0

router2(config-router)#end

router2#

（3）配置路由器 router3，启用 RIP 协议。

Router3(config)#router   rip

Router3(config-router)#version   2

Router3(config-router)#no   auto-summary

Router3(config-router)#network   10.0.0.0

Router3(config-router)#network   192.168.20.0

Router3(config-router)#end

Router3#

步骤 3 观察路由表。

router1#show   ip   route

Codes: C - connected, S - static, I - IGRP, R - RIP, M - mobile, B - BGP

   D - EIGRP, EX - EIGRP external, O - OSPF, IA - OSPF inter area

   N1 - OSPF NSSA external type 1, N2 - OSPF NSSA external type 2

   E1 - OSPF external type 1, E2 - OSPF external type 2, E - EGP

   i - IS-IS, L1 - IS-IS level-1, L2 - IS-IS level-2, ia - IS-IS inter area

   * - candidate default, U - per-user static route, o - ODR

   P - periodic downloaded static route

Gateway of last resort is not set

  10.0.0.0/30 is subnetted, 2 subnets

C 10.1.1.0 is directly connected, FastEthernet0/0

R 10.1.1.4 [120/1] via 10.1.1.2, 00:00:06, FastEthernet0/0

C 192.168.10.0/24 is directly connected, FastEthernet0/1

R 192.168.20.0/24 [120/2] via 10.1.1.2, 00:00:06, FastEthernet0/0

router1#

以上输出表明路由器 router1 学到了 2 条 RIP 路由，其中"R   192.168.20.0/24 [120/2] via 10.1.1.2, 00:00:06, FastEthernet0/0"的含义如下。

①R：该路由条目由 RIP 学习而来。

②192.168.20.0/24：目的网络。

③120：RIP 的默认管理距离为 120。

④2：度量值，从路由器 router1 到达网络 192.168.20.0/24 的度量值是 2 跳。

⑤10.1.1.2：下一跳的 IP 地址。

⑥00:00:06：路由条目 6s 前更新。

⑦FastEthernet0/0：接收该路由条目的接口。

同样，我们可以观察路由器 router2 和 router3 的路由表，请同学们自己解释各路由条目的含义。

router2#show   ip route

Codes: C - connected, S - static, I - IGRP, R - RIP, M - mobile, B - BGP

D - EIGRP, EX - EIGRP external, O - OSPF, IA - OSPF inter area

N1 - OSPF NSSA external type 1, N2 - OSPF NSSA external type 2

E1 - OSPF external type 1, E2 - OSPF external type 2, E - EGP

i - IS-IS, L1 - IS-IS level-1, L2 - IS-IS level-2, ia - IS-IS inter area

* - candidate default, U - per-user static route, o - ODR

P - periodic downloaded static route

Gateway of last resort is not set

  10.0.0.0/30 is subnetted, 2 subnets

C 10.1.1.0 is directly connected, FastEthernet0/0

C 10.1.1.4 is directly connected, Serial0/0/0

R 192.168.10.0/24 [120/1] via 10.1.1.1, 00:00:26, FastEthernet0/0

R 192.168.20.0/24 [120/1] via 10.1.1.6, 00:00:07, Serial0/0/0

router2#

Router3#show ip route

Codes: C - connected, S - static, I - IGRP, R - RIP, M - mobile, B - BGP

   D - EIGRP, EX - EIGRP external, O - OSPF, IA - OSPF inter area

   N1 - OSPF NSSA external type 1, N2 - OSPF NSSA external type 2

   E1 - OSPF external type 1, E2 - OSPF external type 2, E - EGP

   i - IS-IS, L1 - IS-IS level-1, L2 - IS-IS level-2, ia - IS-IS inter area

   * - candidate default, U - per-user static route, o - ODR

   P - periodic downloaded static route

Gateway of last resort is not set

  10.0.0.0/30 is subnetted, 2 subnets

R 10.1.1.0 [120/1] via 10.1.1.5, 00:00:18, Serial0/0/0

C 10.1.1.4 is directly connected, Serial0/0/0

R 192.168.10.0/24 [120/2] via 10.1.1.5, 00:00:18, Serial0/0/0

C 192.168.20.0/24 is directly connected, FastEthernet0/1

Router#

步骤4 验证测试网络连通性。

使用计算机 pc0 去 ping 计算机 pc1：

PC>ping 192.168.20.2

Pinging 192.168.20.2 with 32 bytes of data:

Reply from 192.168.20.2: bytes=32 time=125ms TTL=125

Reply from 192.168.20.2: bytes=32 time=125ms TTL=125

Reply from 192.168.20.2: bytes=32 time=125ms TTL=125

Reply from 192.168.20.2: bytes=32 time=94ms TTL=125

Ping statistics for 192.168.20.2:

    Packets: Sent = 4, Received = 4, Lost = 0 (0% loss),

Approximate round trip times in milli-seconds:

Minimum = 94ms, Maximum = 125ms, Average = 117ms

任务4.2 单区域的 OSPF

【实训目的】

通过配置路由协议 OSPF，理解 OSPF 的路由学习原理，比较其与 RIP 路由静态路由的不同。

【实训任务】

1. 根据拓扑图连接网络设备，构建局域网。
2. 配置路由协议 OSPF，观察其路由表。
3. 测试网络连通性。

【预备知识】

一、OSPF 路由协议

OSPF（开放最短路径优先）路由协议是一个链路状态路由协议。链路即路由器连接网络的接口。OSPF 通过路由器之间通告网络接口的状态来建立链路状态数据库，每个路由器生成以自己为根的最短路径树，然后根据最短路径树构造路由表。

OSPF 路由器向加入到 OSPF 过程的接口发送 Hello 数据包，Hello 协议的目的主要有两个：一是用于发现邻居，建立邻接关系；二是用于在 NBMA（Nonbroadcast Multi-access）网络上选举 DR 和 BDR。

当路由器之间形成邻居关系之后，OSPF 将链路状态数据 LSA（Link State Advertisement）通过组播地址 224.0.0.5（注意：所有 SPF 路由器的组播地址是 224.0.0.5，所有 DR 的组播地址是 224.0.0.6）传送给在某一区域内的所有路由器，这一点与距离矢量路由协议不同。运行距离矢量路由协议的路由器是将部分或全部的路由表传递给与其相邻的路由器。当所有路由器收到 LSA 后运行 Dijkstra SPF 算法来了解到达所有目的地的最短路径，然后创建 SPF 树，再根据 SPF 树的信息创建路由表。OSPF 的管理距离 110，跳数无限大，突破了 RIP 只允许最大 15 跳的限制，适合中大型网络。OSPF 支持等价负载均衡、手动汇总和认证功能。

OSPF 路由协议采用 COST 作为度量标准，COST 的计算公式是 10^8/带宽，然后取整，而且是所有链路入口的 COST 之和。OSPF 路由器根据最短路径优先算法（SPF）各自独立地计算到达每个目的网络的最佳路由。同一区域的所有路由器拥有一个相同的链路状态数据库。OSPF 的管理距离值小于 RIP，当存在到达某一共同网段时，路由器优先选择 OSPF 路由。

二、OSPF 定义了四种网络类型

1. 广播多路访问型（Broadcast multiAccess）
2. 非广播多路访问型（None Broadcast MultiAccess，NBMA）
3. 点到点型（Pointerface-to-Pointerface）
4. 点到多点型（Pointerface-to-MultiPointerface）

三、OSPF 配置命令

Router(config)#router   ospf   进程号

Router(config-router)#router-id   X.X.X.X

Router(config-router)#network   网络号   掩码|反掩码   area   区域号

技术要点如下。

（1）OSPF 进程号，范围为 1~65535，用于在路由器本地标识 OSPF 进程，只具有本地意义，不同路由器的进程号可以不相同。

（2）为了方便管理运行 OSPF 的路由器，我们通常需要手工指定 router-id。如果不指定，router-id 的选择顺序如下所示：

①路由器最大的环回接口的 IP 作为 router-id；

②路由器没有环回接口，则使用活动接口的最大 IP 作为 router-id。

（3）OSPF 是无类路由协议，通告的网络号后须加掩码或反掩码。

（4）区域号可以采用 IP 地址 A.B.C.D 方式表示，也可以用 0~4294967295 的十进制数表示。区域 0 是骨干区域。

【实训拓扑】

网络拓扑结构图如图 4-2 所示。

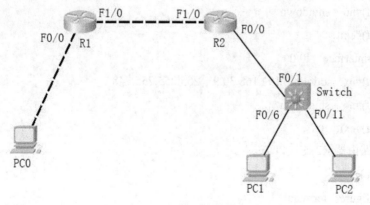

图 4-2 网络拓扑结构图

【实训设备】

路由器 2 台、三层交换机 1 台、计算机 3 台。

【实训步骤】

步骤 1 配置各网络设备的接口 IP，参见表 4-2 所示。

表 4-2

设备名称	接口	IP	子网掩码
R1	F0/0	192.168.30.9	255.255.255.248
	F1/0	192.168.20.1	255.255.255.252
R2	F1/0	192.168.20.2	255.255.255.252
	F0/0	192.168.10.1	255.255.255.224
Switch	VLAN10(F0/1)	192.168.10.2	255.255.255.224
	VLAN20(F0/6)	192.168.10.33	255.255.255.240
	VLAN30(F0/11)	192.168.10.65	255.255.255.192
PC0	网卡	192.168.30.10	255.255.255.248
Pc1	网卡	192.168.10.34	255.255.255.240
Pc2	网卡	192.168.10.66	255.255.255.192

（1）配置路由器 R1 的接口 IP。

Router>enable

Router#configure terminal

Router(config)#hostname R1

R1(config)#interface f1/0

R1(config-if)#ip address 192.168.20.1 255.255.255.252

R1(config-if)#no shutdown

R1(config-if)#exit

R1(config)#interface f0/0

R1(config-if)#ip address 192.168.30.9 255.255.255.248

R1(config-if)#no shutdown

R1(config-if)#exit

（2）配置路由器 R2 的接口 IP。

Router>enable

Router#configure terminal

Router(config)#interface   f0/0

Router(config-if)#ip   address   192.168.10.1   255.255.255.224

Router(config-if)#no   shutdown

Router(config-if)#exit

Router(config)#interface   f1/0

Router(config-if)#ip   address   192.168.20.2   255.255.255.252

Router(config-if)#no   shutdown

Router(config-if)#exit

Router(config)#

Router#configure   terminal

Router(config)#hostname   R2

（3）配置三层交换机。

Switch>

Switch>enable

Switch#configure   terminal

Switch(config)#vlan   10

Switch(config-vlan)#vlan   20

Switch(config-vlan)#vlan   30

Switch(config-vlan)#exit

Switch(config)#interface   f0/1

Switch(config-if)#switchport   access   vlan   10

Switch(config-if)#exit

Switch(config)#interface   f0/6

Switch(config-if)# switchport   access   vlan   20

Switch(config-if)#exit

Switch(config)#interface   f0/11

Switch(config-if)# switchport   access   vlan   30

Switch(config-if)#end

Switch#show   vlan

VLAN Name   Status   Ports
---- -------------------------------- --------- -------------------------------
1   default   active   Fa0/2, Fa0/3, Fa0/4, Fa0/5
                                                  Fa0/7, Fa0/8, Fa0/9, Fa0/10
                                                  Fa0/12, Fa0/13, Fa0/14, Fa0/15
                                                  Fa0/16, Fa0/17, Fa0/18, Fa0/19

			Fa0/20, Fa0/21, Fa0/22, Fa0/23
			Fa0/24, Gig0/1, Gig0/2
10	VLAN0010	active	Fa0/1
20	VLAN0020	active	Fa0/6
30	VLAN0030	active	Fa0/11

Switch#configure terminal

Switch(config)#interface vlan 10

Switch(config-if)#ip address 192.168.10.2 255.255.255.224

Switch(config-if)#no shutdown

Switch(config-if)#exit

Switch(config)#interface vlan 20

Switch(config-if)#ip address 192.168.10.33 255.255.255.240

Switch(config-if)#no shutdown

Switch(config-if)#exit

Switch(config)#interface vlan 30

Switch(config-if)#ip address 192.168.10.65 255.255.255.192

Switch(config-if)#no shutdown

Switch(config-if)#exit

Switch(config)#ip routing //开启交换机的路由

步骤2 配置各设备的路由协议 OSPF。

（1）配置路由器 R1，启用 OSPF。

R1(config)#router ospf 1

R1(config-router)#router-id 1.1.1.1

R1(config-router)#network 192.168.30.8 0.0.0.7 area 0

R1(config-router)#network 192.168.20.0 0.0.0.3 area 0

R1(config-router)#end

R1#

（2）配置路由器 R2，启用 OSPF。

R2(config)#router ospf 1

R2(config-router)#router-id 2.2.2.2

R2(config-router)#network 192.168.20.0 0.0.0.3 area 0

R2(config-router)#network 192.168.10.0 0.0.0.31 area 0

R2(config-router)#end

R2#

（3）配置三层交换机，启用 OSPF。

Switch(config)#router ospf 1

Switch(config-router)#router-id 3.3.3.3
Switch(config-router)#network   192.168.10.0   0.0.0.31   area   0
Switch(config-router)#network   192.168.10.32   0.0.0.15   area   0
Switch(config-router)#network   192.168.10.32   0.0.0.15   area   0
Switch(config-router)#network   192.168.10.64   0.0.0.63   area   0
Switch(config-router)#end
Switch#

步骤3 观察各网络设备的路由表。

R1#show   ip   route
Codes: C - connected, S - static, I - IGRP, R - RIP, M - mobile, B - BGP
　　　　D - EIGRP, EX - EIGRP external, O - OSPF, IA - OSPF inter area
　　　　N1 - OSPF NSSA external type 1, N2 - OSPF NSSA external type 2
　　　　E1 - OSPF external type 1, E2 - OSPF external type 2, E - EGP
　　　　i - IS-IS, L1 - IS-IS level-1, L2 - IS-IS level-2, ia - IS-IS inter area
　　　　* - candidate default, U - per-user static route, o - ODR
　　　　P - periodic downloaded static route

Gateway of last resort is not set

　　　　192.168.10.0/24 is variably subnetted, 3 subnets, 3 masks
O　　　　192.168.10.0/27 [110/2] via 192.168.20.2, 00:02:37, FastEthernet1/0
O　　　　192.168.10.32/28 [110/3] via 192.168.20.2, 00:01:34, FastEthernet1/0
O　　　　192.168.10.64/26 [110/3] via 192.168.20.2, 00:00:43, FastEthernet1/0
　　　　192.168.20.0/30 is subnetted, 1 subnets
C　　　　192.168.20.0 is directly connected, FastEthernet1/0
　　　　192.168.30.0/29 is subnetted, 1 subnets
C　　　　192.168.30.8 is directly connected, FastEthernet0/0

以上输出表明路由器 R1 学到了 3 条 OSPF 路由，其中"O   192.168.10.64/26 [110/3] via 192.168.20.2, 00:00:43, FastEthernet1/0"的含义如下。

①O：该路由条目由 OSPF 学习而来。

②192-168-10-64/26：目的网络。

③120：OSPF 的默认管理距离。

④3：COST 值，三层交换机的 VLAN30 接口的 COST 值是 1，路由器 R2 的 F0/0 的 COST 是 1，路由器 R1 的 F1/0 的 COST 是 1，以上三个入口 COST 值的和就是 3。

⑤192.168.20.2：下一跳地址。

⑥00:00:43：距离下一次更新还有 43s。

⑦FastEthernet1/0：接收该路由条目的接口。

同样，我们可以观察路由器 router2 和 switch 的路由表，请同学们自己解释各路由条目的含义。

R2#show　ip　route

Codes: C - connected, S - static, I - IGRP, R - RIP, M - mobile, B - BGP

  D - EIGRP, EX - EIGRP external, O - OSPF, IA - OSPF inter area

  N1 - OSPF NSSA external type 1, N2 - OSPF NSSA external type 2

  E1 - OSPF external type 1, E2 - OSPF external type 2, E - EGP

  i - IS-IS, L1 - IS-IS level-1, L2 - IS-IS level-2, ia - IS-IS inter area

  * - candidate default, U - per-user static route, o - ODR

  P - periodic downloaded static route

Gateway of last resort is not set

  192.168.10.0/24 is variably subnetted, 3 subnets, 3 masks

C 192.168.10.0/27 is directly connected, FastEthernet0/0

O 192.168.10.32/28 [110/2] via 192.168.10.2, 00:01:22, FastEthernet0/0

O 192.168.10.64/26 [110/2] via 192.168.10.2, 00:00:32, FastEthernet0/0

  192.168.20.0/30 is subnetted, 1 subnets

C 192.168.20.0 is directly connected, FastEthernet1/0

  192.168.30.0/29 is subnetted, 1 subnets

O 192.168.30.8 [110/2] via 192.168.20.1, 00:02:25, FastEthernet1/0

Switch#show　ip　route

Codes: C - connected, S - static, I - IGRP, R - RIP, M - mobile, B - BGP

  D - EIGRP, EX - EIGRP external, O - OSPF, IA - OSPF inter area

  N1 - OSPF NSSA external type 1, N2 - OSPF NSSA external type 2

  E1 - OSPF external type 1, E2 - OSPF external type 2, E - EGP

  i - IS-IS, L1 - IS-IS level-1, L2 - IS-IS level-2, ia - IS-IS inter area

  * - candidate default, U - per-user static route, o - ODR

  P - periodic downloaded static route

Gateway of last resort is not set

  192.168.10.0/24 is variably subnetted, 3 subnets, 3 masks

C 192.168.10.0/27 is directly connected, Vlan10

C 192.168.10.32/28 is directly connected, Vlan20

C 192.168.10.64/26 is directly connected, Vlan30

  192.168.20.0/30 is subnetted, 1 subnets

O 192.168.20.0 [110/2] via 192.168.10.1, 00:20:44, Vlan10

  192.168.30.0/29 is subnetted, 1 subnets

O 192.168.30.8 [110/3] via 192.168.10.1, 00:20:44, Vlan10

步骤4 查看路由器 R2 的邻居。

r2#show　ip　ospf　neighbor

Neighbor ID	Pri	State	Dead Time	Address	Interface
3.3.3.3	1	FULL/BDR	00:00:39	192.168.10.2	FastEthernet0/0
1.1.1.1	1	FULL/BDR	00:00:39	192.168.20.1	FastEthernet1/0

以上输出表明路由器 R2 有两个邻居，它们的路由器 ID 分别是 1.1.1.1 和 3.3.3.3,其他参数含义如下。

①Pri：邻居路由器接口的优先级，如果为 0，则无权参加 DR 的选举。

②State：当前路由器接口的状态。

③Dead Time：清除邻居关系前需等待的最长时间。

④Adress：邻居接口的地址。

⑤Interface：设备本身和邻居相连的接口。

⑥BDR:表示邻居路由器是备份指定路由器。

邻居不能形成的常见原因如下。

①HELLO 间隔和 DEAD 间隔不一致。

②区域类型不匹配。

③认证类型或密码不一致。

④HELLO 包被访问列表拒绝。

⑤链路上的 MTU 不相等。

⑥接口下 OSPF 的网络类型不匹配。

其他常使用的命令有 show ip ospf database、show ip ospf interface、show ip protocols 等，请同学们自己尝试运行，分析其用途。

步骤 5 测试网络连通性。

测试计算机 pc0 和计算机 pc1 能否互通。

PC>ping   192.168.10.34

Pinging 192.168.10.34 with 32 bytes of data:

Reply from 192.168.10.34: bytes=32 time=125ms TTL=125

Reply from 192.168.10.34: bytes=32 time=125ms TTL=125

Reply from 192.168.10.34: bytes=32 time=125ms TTL=125

Reply from 192.168.10.34: bytes=32 time=125ms TTL=125

Ping statistics for 192.168.10.34:

　　Packets: Sent = 4, Received = 4, Lost = 0 (0% loss),

Approximate round trip times in milli-seconds:

Minimum = 125ms, Maximum = 125ms, Average = 125ms

以上输出表明计算机 pc0 和计算机 pc1 能够互通。请同学们自己测试其他设备之间的互通性。

任务 4.3 OSPF 基于区域的 MD5 认证

【实训目的】

配置基于区域的 OSPF MD5 认证，提高网络安全性。

【实训任务】

1. 配置 OSPF 路由协议。
2. 配置基于区域的 OSPF MD5 认证。
3. 验证测试 OSPF MD5 认证。

【预备知识】

OSPF 路由协议支持认证，有效阻止非法网络设备的接入，大大提高了网络的安全性。OSPF 认证分为区域认证和接口认证，加密方法分为明文和 MD5 加密。

一、域认证配置命令

Router(config)#router ospf 进程号
Router (config-router)#area 区域号 authentication [message-digest]
　// 选择 essage-digest 选项，加密方式为 MD5；省略 message-digest 选项，则加密方式为明文。
Router (config)#interface 接口名称
Router (config-if)#ip ospf message-digest-key 1 md5 密码//在接口下应用认证，并设置验证密码。

二、口认证配置命令

Router (config)#interface 接口名称
Router (config-if)#ip ospf authentication [message-digest]
Router (config-if)#ip ospf message-digest-key 1 md5 密码

【实训拓扑】

网络拓扑结构图如图 4-3 所示

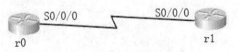

图 4-3　网络拓扑结构图

【实训设备】

路由器 2 台。

【实训步骤】

步骤 1 配置路由器名称 r0 和 r1 接口 IP 地址，参见表 4-3 所示。

表 4-3

设备名称	接口	IP 地址	子网掩码
r0	Loopback0	1.1.1.1	255.255.255.255
	S0/0/0	12.1.1.1	255.255.255.252
r1	Loopback0	2.2.2.2	255.255.255.255
	S0/0/0	12.1.1.2	255.255.255.252

（1）配置路由器名称 r0 和接口 IP 地址。

Router>enable

Router#configure terminal

Router(config)#hostname r0

r0(config)#interface s0/0/0

r0(config-if)#ip address 12.1.1.1 255.255.255.0

r0(config-if)#clock rate 128000

r0(config-if)#no shutdown

r0(config-if)#exit

r0(config)#interface lo0

r0(config-if)#ip address 1.1.1.1 255.255.255.255

r0(config-if)#exit

r0(config)#

（2）配置路由器名称 r1 和接口 IP 地址。

Router>enable

Router#configrue terminal

Router(config)#hostname r1

r1(config)#interface s0/0/0

r1(config-if)#ip address 12.1.1.2 255.255.255.0

r1(config-if)#no shutdown

r1(config-if)#exit

r1(config)#interface lo0

r1(config-if)#ip address 2.2.2.2 255.255.255.255

r1(config-if)#exit

r1(config)#

步骤 2 测试连通性。

r0#ping 12.1.1.2

Type escape sequence to abort.

Sending 5, 100-byte ICMP Echos to 12.1.1.2, timeout is 2 seconds:

!!!!!

Success rate is 100 percent (5/5), round-trip min/avg/max = 31/31/32 ms

r0#ping 2.2.2.2

Type escape sequence to abort.

Sending 5, 100-byte ICMP Echos to 2.2.2.2, timeout is 2 seconds:

.....

Success rate is 0 percent (0/5)

为什么路由器 r0 可以 ping 通 12.1.1.2，却 ping 不通 2.2.2.2？观察 r0 的路由表：

r0#show ip route

Codes: C - connected, S - static, I - IGRP, R - RIP, M - mobile, B - BGP

   D - EIGRP, EX - EIGRP external, O - OSPF, IA - OSPF inter area

   N1 - OSPF NSSA external type 1, N2 - OSPF NSSA external type 2

   E1 - OSPF external type 1, E2 - OSPF external type 2, E - EGP

   i - IS-IS, L1 - IS-IS level-1, L2 - IS-IS level-2, ia - IS-IS inter area

   * - candidate default, U - per-user static route, o - ODR

   P - periodic downloaded static route

Gateway of last resort is not set

  1.0.0.0/32 is subnetted, 1 subnets

C 1.1.1.1 is directly connected, Loopback0

  12.0.0.0/24 is subnetted, 1 subnets

C 12.1.1.0 is directly connected, Serial0/0/0

仔细观察，发现路由器 r0 上有 12.1.1.0 的直连路由，而不存在到达 2.2.2.2 的路由。然后查看 r1 的路由表：

r1#show ip route

Codes: C - connected, S - static, I - IGRP, R - RIP, M - mobile, B - BGP

   D - EIGRP, EX - EIGRP external, O - OSPF, IA - OSPF inter area

N1 - OSPF NSSA external type 1, N2 - OSPF NSSA external type 2

E1 - OSPF external type 1, E2 - OSPF external type 2, E - EGP

i - IS-IS, L1 - IS-IS level-1, L2 - IS-IS level-2, ia - IS-IS inter area

* - candidate default, U - per-user static route, o - ODR

P - periodic downloaded static route

Gateway of last resort is not set

  2.0.0.0/32 is subnetted, 1 subnets

C 2.2.2.2 is directly connected, Loopback0

  12.0.0.0/24 is subnetted, 1 subnets

C 12.1.1.0 is directly connected, Serial0/0/0

由此发现路由器 r1 上也存在到达 12.1.1.0 的直连路由。所以路由器 r0 可以 ping 通 12.1.1.2，却 ping 不通 2.2.2.2。

步骤 3　在路由器 r0 和 r1 上分别配置 OSPF 协议。

（1）在路由器 r0 上配置 OSPF 协议。

r0(config)#router　ospf　1

r0(config-router)#router-id　1.1.1.1

r0(config-router)#network　12.1.1.0　0.0.0.255　area　0

r0(config-router)#network　1.1.1.1　0.0.0.0　area　0

r0(config-router)#end

（2）在路由器 r1 上配置 OSPF 协议。

r1(config)#router　ospf　1

r1(config-router)#router-id　2.2.2.2

r1(config-router)#network　12.1.1.0　0.0.0.255　area　0

r1(config-router)#network　2.2.2.2　0.0.0.0　area　0

r1(config-router)#end

步骤 4　查看路由器 r0 和 r1 的路由表。

（1）查看路由器 r0 的路由表。

r0#show ip route

Codes: C - connected, S - static, I - IGRP, R - RIP, M - mobile, B - BGP

  D - EIGRP, EX - EIGRP external, O - OSPF, IA - OSPF inter area

  N1 - OSPF NSSA external type 1, N2 - OSPF NSSA external type 2

  E1 - OSPF external type 1, E2 - OSPF external type 2, E - EGP

  i - IS-IS, L1 - IS-IS level-1, L2 - IS-IS level-2, ia - IS-IS inter area

  * - candidate default, U - per-user static route, o - ODR

  P - periodic downloaded static route

Gateway of last resort is not set

　　　　　1.0.0.0/32 is subnetted, 1 subnets

C　　　　1.1.1.1 is directly connected, Loopback0

　　　　　2.0.0.0/32 is subnetted, 1 subnets

O　　　　2.2.2.2 [110/65] via 12.1.1.2, 00:04:00, Serial0/0/0

　　　　　12.0.0.0/24 is subnetted, 1 subnets

C　　　　12.1.1.0 is directly connected, Serial0/0/0

结果显示：路由器 r0 学习到了路由器 r1 环回口路由 "O　　　　2.2.2.2 [110/65] via 12.1.1.2, 00:04:00, Serial0/0/0"。

（2）查看路由器 r1 的路由表。

r1#show ip route

Codes: C - connected, S - static, I - IGRP, R - RIP, M - mobile, B - BGP

　　　　D - EIGRP, EX - EIGRP external, O - OSPF, IA - OSPF inter area

　　　　N1 - OSPF NSSA external type 1, N2 - OSPF NSSA external type 2

　　　　E1 - OSPF external type 1, E2 - OSPF external type 2, E - EGP

　　　　i - IS-IS, L1 - IS-IS level-1, L2 - IS-IS level-2, ia - IS-IS inter area

　　　　* - candidate default, U - per-user static route, o - ODR

　　　　P - periodic downloaded static route

Gateway of last resort is not set

　　　　　1.0.0.0/32 is subnetted, 1 subnets

O　　　　1.1.1.1 [110/65] via 12.1.1.1, 00:04:23, Serial0/0/0

　　　　　2.0.0.0/32 is subnetted, 1 subnets

C　　　　2.2.2.2 is directly connected, Loopback0

　　　　　12.0.0.0/24 is subnetted, 1 subnets

C　　　　12.1.1.0 is directly connected, Serial0/0/0

路由器 r1 学习到了路由器 r0 环回口路由 "O　　　　1.1.1.1 [110/65] via 12.1.1.1, 00:04:23, Serial0/0/0"。

步骤 5　测试连通性。

r0#ping　2.2.2.2

Type escape sequence to abort.

Sending 5, 100-byte ICMP Echos to 2.2.2.2, timeout is 2 seconds:

!!!!!

Success rate is 100 percent (5/5), round-trip min/avg/max = 16/28/31 ms

现在，路由器 r0 可以 ping 通路由器 r1 的环回口 2.2.2.2 了。

步骤 6　观察路由器 r0 的邻居。

r0#show　ip　ospf　neighbor

Neighbor ID　　　Pri　　State　　　　　Dead Time　　Address　　　　Interface

2.2.2.2   0   FULL/ -   00:00:37   12.1.1.2   Serial0/0/0

步骤 7 配置基于区域的 OSPF MD5 认证。

（1）配置路由器 r0 基于区域的 OSPF MD5 认证。

r0#configure terminal

r0(config)#router ospf 1

r0(config-router)#area 0 authentication message-digest //区域 0 验证方式为 MD5

r0(config)#interface s0/0/0

r0(config-if)#ip ospf message-digest-key 1 md5 cisco //验证密码为 cisco

r0(config-if)#

配置路由器 r0 的 MD5 认证后，系统会出现如下提示：

00:55:32: %OSPF-5-ADJCHG: Process 1, Nbr 2.2.2.2 on Serial0/0/0 from FULL to DOWN, Neighbor Down: Dead timer expired

00:55:32: %OSPF-5-ADJCHG: Process 1, Nbr 2.2.2.2 on Serial0/0/0 from FULL to Down: Interface down or detached

这是因为路由器 r1 尚未配置认证，故而提示在接口 Serial0/0/0 上的邻居 2.2.2.2 断开了。

再观察路由器 r0 的邻居，发现确实没有邻居存在了。

r0#show ip ospf neighbor

r0#

（2）配置路由器 r1 基于区域的 OSPF MD5 认证。

r1#configure terminal

r1(config)#router ospf 1

r1(config-router)#area 0 authentication message-digest //区域 0 验证方式为 MD5

r1(config)#interface s0/0/0

r1(config-if)#ip ospf message-digest-key 1 md5 cisco // 验证密码为 cisco

r1(config-if)#

在路由器 r1 的 s0/0/0 上配置 MD5 认证后，提示和路由器 r0 的邻接关系恢复了。

01:02:51: %OSPF-5-ADJCHG: Process 1, Nbr 1.1.1.1 on Serial0/0/0 from EXCHANGE to FULL, Exchange Done

再观察 r0 的邻居，又可以看到邻居路由器 r1 了。

r0#show ip ospf neighbor

Neighbor ID	Pri	State	Dead Time	Address	Interface
2.2.2.2	0	FULL/ -	00:00:35	12.1.1.2	Serial0/0/0

r0#

想一想：配置 OSPF MD5 认证后，如果有外来运行 OSPF 路由协议的非法路由器接入网络，会不会被成功阻止呢？请自己测试一下。

任务 4.4　多区域 OSPF

【实训目的】

了解 OSPF 协议 LSA 的类型，掌握末节区域、完全末节区域和次末节区域的配置。

【实训任务】

1. 配置多区域 OSPF。
2. 查看 OSPF 链路状态数据库。
3. 配置末节区域。
4. 配置完全末节区域。
5. 配置次末节区域。

【预备知识】

一、多区域的好处

1. 减少链路状态数据库的大小，降低 SPF 计算频率。
2. 提高路由的效率：缩减部分路由器的 OSPF 路由条目，降低路由收敛的复杂度，对某些特定的 LSA，可以在区域边界上实现汇总/过滤/控制，从而实现全网互通。
3. 提高网络的稳定性：将不稳定限制在特定的区域。

二、OSPF 中路由器的分类

1. 内部路由器：是指所有接口都在一个区域的路由器。
2. 区域边界路由器(ABR)：是指连接一个或多个区域到骨干区域的路由器，并且这些路由器会作为域间通信量的路由网关。ABR 路由器总是至少有一个接口是属于骨干区域的。
3. 自治系统边界路由器（ASBR）：是 OSPF 域外部的通信量进入 OSPF 域的网关路由器。

三、链路状态通告(LSA)类型（见表 4-4-1）

表 4-4-1

LSA 类型	名称	描述
1	路由器 LSA（O——OSPF）	由各路由器为它的所属区域而生成，描述路由器到该区域链路的状态
2	网络 LSA（O——OSPF）	由 DR 产生，出现在多路访问型网络中

续表

LSA 类型	名称	描述
3	网络汇总 LSA（O IA——OSPF 区域间）	由 ABR 产生，描述 ABR 和本地区域内部路由器之间的链路
4	ASB 汇总 LSA（O IA——OSPF 区域间）	由 ABR 产生，描述如何到达 ASBR
5	自治系统外部 LSA(O E1——OSPF 外部类型 1；O E2——OSPF 外部类型 2)	由 ASBR 产生，描述到自治系统外部目的地的路由
7	自治系统外部链路条目（O N1——OSPF NSSA 外部类型 1；O N2——OSPF 外部类型 2）	由一个连接到 NSSA 的 ASBR 产生，在 NSSA 区域扩散。ABR 可以将类型 7 的 LSA 转化为类型 5 的 LSA

四、区域类型

1. 标准区域：能够接收链路更新和路由汇总。

2. 主干区域：互联多个区域时，该区域为所有其他区域的中心实体，标注为区域 0。

3. 末节区域（Stub Area）：它接收本自治系统（OSPF 网络）中其他区域的路由汇总，拒绝本自治系统以外的路由信息。

4. 完全末节区域：不接收外部自治系统路由及本系统其他区域的路由汇总，向外通信使用默认路由（0.0.0.0/0）。

5. 次末节区域（Not-so-stubby area，NSSA）：与末节区域较为相似，但它只接收类型 7 LSA 的外部路由信息。并且 ABR 负责将类型 7 的 LSA 转化为类型 5 的 LSA。

【实训拓扑】

网络拓扑结构图如图 4-4 所示

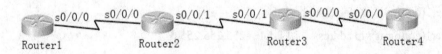

图 4-4　网络拓扑结构图

【实训设备】

路由器 4 台。

【实训步骤】

步骤 1　根据拓扑图构建网络，并配置各路由器接口 IP，参见表 4-4-2 所示。

表 4-4-2

设备名称	接口	IP	子网掩码
Router1	S0/0/0	12.1.1.1	255.255.255.252
	Loopback1	10.1.1.1	255.255.255.0
	Loopback2	10.1.2.1	255.255.255.0
	Loopback3	10.1.3.1	255.255.255.0
Router2	S0/0/0	12.1.1.2	255.255.255.252
	S0/0/1	23.1.1.1	255.255.255.252
Router3	S0/0/1	23.1.1.2	255.255.255.252
	S0/0/0	34.1.1.1	255.255.255.252
Router4	S0/0/0	34.1.1.2	255.255.255.252
	Loopback1	192.168.1.1	255.255.255.0

（1）配置 Router1。

Router>enable

Router#configure terminal

Router(config)#hostname Router1

Router1(config)#interface s0/0/0

Router1(config-if)#ip address 12.1.1.1   255.255.255.252

Router1(config-if)#clock rate   64000

Router1(config-if)#no shutdown

Router1(config-if)#exit

Router1(config)#interface loopback 1

Router1(config-if)#ip address   10.1.1.1   255.255.255.0

Router1(config-if)#exit

Router1(config)#interface loopback 2

Router1(config-if)#ip address   10.1.2.1   255.255.255.0

Router1(config-if)#exit

Router1(config)#interface loopback 3

Router1(config-if)#ip address   10.1.3.1   255.255.255.0

Router1(config-if)#exit

Router1(config)#

（2）配置 Router2。

Router>enable

Router#configure terminal

Router(config)#hostname Router2

Router2(config)#interface s0/0/0

Router2(config-if)#ip address   12.1.1.2   255.255.255.252

Router2(config-if)#no shutdown

Router2(config-if)#exit

Router2(config)#interface s0/0/1

Router2(config-if)#ip address 23.1.1.1   255.255.255.252

Router2(config-if)#clock   rate   64000

Router2(config-if)#no shutdown

Router2(config-if)#exit

Router2(config)#

（3）配置 Router3。

Router>enable

Router#configure terminal

Router(config)#hostname   Router3

Router3(config)#interface s0/0/1

Router3(config-if)#ip   address   23.1.1.2 255.255.255.252

Router3(config-if)#no   shutdown

Router3(config-if)#exit

Router3(config)#

Router3(config)#interface   s0/0/0

Router3(config-if)#ip address   34.1.1.1   255.255.255.252

Router3(config-if)#clock   rate   64000

Router3(config-if)#no shutdown

Router3(config-if)#exit

Router3(config)#

（4）配置 Router4。

Router>enable

Router#configure terminal

Router(config)#hostname   Router4

Router4(config)#interface s0/0/0

Router4(config-if)#ip address   34.1.1.2   255.255.255.252

Router4(config-if)#no   shutdown

Router4(config-if)#exit

Router4(config)#interface   loopback 1

Router4(config-if)#ip address   192.168.1.1   255.255.255.0

Router4(config-if)#exit

Router4(config)#

步骤 2 配置各路由器 OSPF。

(1) 配置 Router1。

Router1(config)#router ospf 1

Router1(config-router)#router-id 1.1.1.1

Router1(config-router)#network 12.1.1.0 0.0.0.3 area 1

Router1(config-router)#network 10.1.1.0 0.0.0.255 area 1

Router1(config-router)#network 10.1.2.0 0.0.0.255 area 1

Router1(config-router)#network 10.1.3.0 0.0.0.255 area 1

Router1(config-router)#end

Router1#

(2) 配置 Router2。

Router2(config)#router ospf 1

Router2(config-router)#router-id 2.2.2.2

Router2(config-router)#network 12.1.1.0 0.0.0.3 area 1

Router2(config-router)#network 23.1.1.0 0.0.0.3 area 0

Router2(config-router)#end

Router2#

(3) 配置 Router3。

Router3(config-router)#router-id 3.3.3.3

Router3(config-router)#network 23.1.1.0 0.0.0.3 area 0

Router3(config-router)#network 34.1.1.0 0.0.0.3 area 2

Router3(config-router)#end

Router3#

(4) 配置 Router4。

Router4(config-router)#router-id 4.4.4.4

Router4(config-router)#network 34.1.1.0 0.0.0.3 area 2

Router4(config-router)#redistribute connected subnets   //重发布直连路由

Router4(config-router)#end

Router4#

步骤 3 查看 OSPF 路由。

(1) 查看 Router1 的 OSPF 路由。

Router1#show ip route ospf

    23.0.0.0/30 is subnetted, 1 subnets

O IA   23.1.1.0 [110/128] via 12.1.1.2, 00:00:16, Serial0/0/0

34.0.0.0/30 is subnetted, 1 subnets
O IA   34.1.1.0 [110/192] via 12.1.1.2, 00:00:06, Serial0/0/0
O E2 192.168.1.0 [110/20] via 12.1.1.2, 00:00:16, Serial0/0/0

（2）查看 Router2 的 OSPF 路由。

Router2#show ip route ospf
    10.0.0.0/32 is subnetted, 3 subnets
O   10.1.1.1 [110/65] via 12.1.1.1, 00:02:12, Serial0/0/0
O   10.1.2.1 [110/65] via 12.1.1.1, 00:02:12, Serial0/0/0
O   10.1.3.1 [110/65] via 12.1.1.1, 00:02:12, Serial0/0/0
    34.0.0.0/30 is subnetted, 1 subnets
O IA   34.1.1.0 [110/128] via 23.1.1.2, 00:02:07, Serial0/0/1
O E2 192.168.1.0 [110/20] via 23.1.1.2, 00:02:07, Serial0/0/1

步骤 4 查看 OSPF 链路状态数据库。

Router1#show ip ospf database
        OSPF Router with ID (1.1.1.1) (Process ID 1)

            Router Link States (Area 1)   //区域 1 类型 1 的 LSA

Link ID	ADV Router	Age	Seq#	Checksum	Link count
1.1.1.1	1.1.1.1	207	0x80000005	0x000b30	5
2.2.2.2	2.2.2.2	207	0x80000002	0x004a4b	2

            Summary Net Link States (Area 1)   //区域 1 类型 3 的 LSA

Link ID	ADV Router	Age	Seq#	Checksum
23.1.1.0	2.2.2.2	212	0x80000001	0x007a86
34.1.1.0	2.2.2.2	192	0x80000003	0x00694a

            Summary ASB Link States (Area 1)   //区域 1 类型 4 的 LSA

Link ID	ADV Router	Age	Seq#	Checksum
4.4.4.4	2.2.2.2	202	0x80000002	0x0007fd

            Type-5 AS External Link States   //类型 5 的 LSA

Link ID	ADV Router	Age	Seq#	Checksum	Tag
192.168.1.0	4.4.4.4	217	0x80000001	0x0094b1	0

Router2#show ip ospf database

OSPF Router with ID (2.2.2.2) (Process ID 1)

Router Link States (Area 0)　　//区域 0 类型 1 的 LSA

Link ID	ADV Router	Age	Seq#	Checksum	Link count
2.2.2.2	2.2.2.2	284	0x80000003	0x00c9ad	2
3.3.3.3	3.3.3.3	284	0x80000003	0x006909	2

Summary Net Link States (Area 0)　　//区域 0 类型 3 的 LSA

Link ID	ADV Router	Age	Seq#	Checksum
12.1.1.0	2.2.2.2	279	0x80000001	0x000a02
34.1.1.0	3.3.3.3	279	0x80000002	0x00cb25
10.1.1.1	2.2.2.2	274	0x80000002	0x0034d3
10.1.2.1	2.2.2.2	274	0x80000003	0x0027de
10.1.3.1	2.2.2.2	274	0x80000004	0x001ae9

Summary ASB Link States (Area 0)　　//区域 0 类型 4 的 LSA

Link ID	ADV Router	Age	Seq#	Checksum
4.4.4.4	3.3.3.3	279	0x80000001	0x00ea17

Router Link States (Area 1)　　//区域 1 类型 1 的 LSA

Link ID	ADV Router	Age	Seq#	Checksum	Link count
2.2.2.2	2.2.2.2	284	0x80000002	0x004a4b	2
1.1.1.1	1.1.1.1	284	0x80000005	0x000b30	5

Summary Net Link States (Area 1)　　//区域 1 类型 3 的 LSA

Link ID	ADV Router	Age	Seq#	Checksum
23.1.1.0	2.2.2.2	289	0x80000001	0x007a86
34.1.1.0	2.2.2.2	269	0x80000003	0x00694a

Summary ASB Link States (Area 1)　　//区域 1 类型 4 的 LSA

Link ID	ADV Router	Age	Seq#	Checksum
4.4.4.4	2.2.2.2	279	0x80000002	0x0007fd

Type-5 AS External Link States　　//类型 5 的 LSA

Link ID	ADV Router	Age	Seq#	Checksum Tag
192.168.1.0	4.4.4.4	294	0x80000001	0x0094b1 0

请观察 Router2 的 OSPF 拓扑结构数据库区域 1 的链路状态，是不是和 Router1 的显示结果一样？事实证明同一区域的链路状态数据库是完全相同的。

步骤 5 配置 area 1 为 STUB 区域。

（1）配置 Router1。

Router1(config)#router   ospf   1

Router1(config-router)#area   1   stub   //配置 area 1 为 STUB 区域

Router1(config-router)#

03:34:42: %OSPF-5-ADJCHG: Process 1, Nbr 2.2.2.2 on Serial0/0/0 from FULL to DOWN, Neighbor Down: Adjacency forced to reset

03:34:42: %OSPF-5-ADJCHG: Process 1, Nbr 2.2.2.2 on Serial0/0/0 from FULL to DOWN, Neighbor Down: Interface down or detached

上述提示信息表示：由于 area 1 上的路由器 Router1 接口 S0/0/0 区域类型为 stub，Router2 接口 S0/0/0 区域类型为标准区域，二者区域类型不同，所以 Router1 和 Router2 无法成为邻居，邻接关系就自然消失了。

（2）配置 Router2。

Router2(config)#router   ospf   1

Router2(config-router)#area   1   stub

Router2(config-router)#end

Router2#

03:35:21: %OSPF-5-ADJCHG: Process 1, Nbr 1.1.1.1 on Serial0/0/0 from LOADING to FULL, Loading Done

//区域 1 上类型一致后，Router2 和 Router1 重新建立了邻接关系。

（3）查看 Router1 的 OSPF 路由、链路状态数据和 Router2 的链路状态数据库。

Router1#show   ip   route   ospf

   23.0.0.0/30 is subnetted, 1 subnets

O IA 23.1.1.0 [110/128] via 12.1.1.2, 00:00:42, Serial0/0/0

   34.0.0.0/30 is subnetted, 1 subnets

O IA 34.1.1.0 [110/192] via 12.1.1.2, 00:00:42, Serial0/0/0

O*IA 0.0.0.0/0 [110/65] via 12.1.1.2, 00:00:42, Serial0/0/0

Router1#show   ip   ospf   database

    OSPF Router with ID (1.1.1.1) (Process ID 1)

    Router Link States (Area 1)

Link ID	ADV Router	Age	Seq#	Checksum Link count
2.2.2.2	2.2.2.2	223	0x80000007	0x004050 2
1.1.1.1	1.1.1.1	881	0x80000007	0x000732 5

Summary Net Link States (Area 1)

Link ID	ADV Router	Age	Seq#	Checksum
23.1.1.0	2.2.2.2	1564	0x80000001	0x007a86
34.1.1.0	2.2.2.2	1544	0x80000003	0x00694a
0.0.0.0	2.2.2.2	876	0x80000004	0x005102

Summary ASB Link States (Area 1)

Link ID	ADV Router	Age	Seq#	Checksum
4.4.4.4	2.2.2.2	1554	0x80000002	0x0007fd

//Router1 的区域 1 没有类型 5 的 LSA

Router2#show ip ospf data

OSPF Router with ID (2.2.2.2) (Process ID 1)

Router Link States (Area 0)

Link ID	ADV Router	Age	Seq#	Checksum Link count
2.2.2.2	2.2.2.2	868	0x80000003	0x00c9ad 2
3.3.3.3	3.3.3.3	868	0x80000003	0x006909 2

Summary Net Link States (Area 0)

Link ID	ADV Router	Age	Seq#	Checksum
12.1.1.0	2.2.2.2	863	0x80000001	0x000a02
34.1.1.0	3.3.3.3	863	0x80000002	0x00cb25
10.1.1.1	2.2.2.2	858	0x80000002	0x0034d3
10.1.2.1	2.2.2.2	858	0x80000003	0x0027de
10.1.3.1	2.2.2.2	858	0x80000004	0x001ae9

Summary ASB Link States (Area 0)

Link ID	ADV Router	Age	Seq#	Checksum
4.4.4.4	3.3.3.3	863	0x80000001	0x00ea17

Router Link States (Area 1)

Link ID	ADV Router	Age	Seq#	Checksum	Link count
1.1.1.1	1.1.1.1	190	0x80000007	0x000732	5
2.2.2.2	2.2.2.2	190	0x80000004	0x00464d	2

Summary Net Link States (Area 1)

Link ID	ADV Router	Age	Seq#	Checksum
23.1.1.0	2.2.2.2	873	0x80000001	0x007a86
34.1.1.0	2.2.2.2	853	0x80000003	0x00694a
0.0.0.0	2.2.2.2	185	0x80000004	0x005102

//Router2 向区域内通告 "0.0.0.0"

Summary ASB Link States (Area 1)

Link ID	ADV Router	Age	Seq#	Checksum
4.4.4.4	2.2.2.2	863	0x80000002	0x0007fd

Type-5 AS External Link States

Link ID	ADV Router	Age	Seq#	Checksum	Tag
192.168.1.0	4.4.4.4	878	0x80000001	0x0094b1	0

以上结果表明：stub 区域可以接收区域间的路由，但不会接收外部路由。并且 Stub 区域和骨干区域 0 交界的路由器 Router2 自动向 stub 传播了一条默认路由。

步骤 6　配置 Area 1 为完全末节区域。

（1）配置 ABR　Router2。

Router2(config)#router　ospf 1

Router2(config-router)#area　1　stub no-summary

Router2(config-router)#end

Router2#

（2）查看 Router1 上的路由。

Router1#show　ip　route

Codes: C - connected, S - static, I - IGRP, R - RIP, M - mobile, B - BGP

　　　　D - EIGRP, EX - EIGRP external, O - OSPF, IA - OSPF inter area

　　　　N1 - OSPF NSSA external type 1, N2 - OSPF NSSA external type 2

　　　　E1 - OSPF external type 1, E2 - OSPF external type 2, E - EGP

　　　　i - IS-IS, L1 - IS-IS level-1, L2 - IS-IS level-2, ia - IS-IS inter area

　　　　* - candidate default, U - per-user static route, o - ODR

　　　　P - periodic downloaded static route

Gateway of last resort is 12.1.1.2 to network 0.0.0.0

    10.0.0.0/24 is subnetted, 3 subnets

C   10.1.1.0 is directly connected, Loopback1

C   10.1.2.0 is directly connected, Loopback2

C   10.1.3.0 is directly connected, Loopback3

    12.0.0.0/30 is subnetted, 1 subnets

C   12.1.1.0 is directly connected, Serial0/0/0

O*IA 0.0.0.0/0 [110/65] via 12.1.1.2, 00:04:20, Serial0/0/0

以上输出表明：

① 完全末节区域只需在区域边界路由器增添"no- summary"选项；

② 完全末节区域只有内部路由条目和 ABR 注入的一条默认路由，不会接收外部路由和区域间的路由。

步骤 7 配置 Area 2 为 NSSA 区域。

（1）配置 Router2。

Router2(config)#interface loopback 1 //开启环回接口，模拟外部路由

Router2(config-if)#ip address 100.1.1.1 255.255.255.0

Router2(config-if)#exit

Router2(config)#router ospf 1

Router2(config-router)#redistribute connected subnets //重发布直连路由

Router2(config-router)#end

Router2#

（2）配置 Router4。

Router4(config)#router ospf 1

Router4(config-router)#area 2 nssa

Router4(config-router)#

00:01:15: %OSPF-5-ADJCHG: Process 1, Nbr 3.3.3.3 on Serial0/0/0 from FULL to DOWN, Neighbor Down: Adjacency forced to reset

00:01:15: %OSPF-5-ADJCHG: Process 1, Nbr 3.3.3.3 on Serial0/0/0 from FULL to DOWN, Neighbor Down: Interface down or detached

//由于 area 2 上的路由器 Router4 接口 S0/0/0 区域类型为 nssa，Router3 接口 S0/0/0 区域类型为标准区域，二者区域类型不同，所示 Router3 和 Router4 无法成为邻居，邻接关系消失。

Router4(config-router)#end

Router4#

（3）配置 Router3。

Router3(config)#router ospf 1

Router3(config-router)#area   2   nssa

00:06:10: %OSPF-5-ADJCHG: Process 1, Nbr 4.4.4.4 on Serial0/0/0 from LOADING to FULL, Loading Done

//区域 1 上类型一致后，Router2 和 Router1 重新建立了邻接关系。

Router3(config-router)#end

Router3#

（4）查看 Router3 和 Router4 上的 OSPF 路由。

Router3#show   ip   route   ospf

  10.0.0.0/32 is subnetted, 3 subnets

O IA 10.1.1.1 [110/129] via 23.1.1.1, 01:30:33, Serial0/0/1

O IA 10.1.2.1 [110/129] via 23.1.1.1, 01:30:33, Serial0/0/1

O IA 10.1.3.1 [110/129] via 23.1.1.1, 01:30:33, Serial0/0/1

  12.0.0.0/30 is subnetted, 1 subnets

O IA 12.1.1.0 [110/128] via 23.1.1.1, 01:30:33, Serial0/0/1

  100.0.0.0/24 is subnetted, 1 subnets

O E2 100.1.1.0 [110/20] via 23.1.1.1, 01:15:22, Serial0/0/1

O N2 192.168.1.0 [110/20] via 34.1.1.2, 01:24:38, Serial0/0/0

//7 类 LSA 在 nssa 区域扩散，形成"O N2"路由

Router4#show   ip   route   ospf

  10.0.0.0/32 is subnetted, 3 subnets

O IA 10.1.1.1 [110/193] via 34.1.1.1, 00:15:24, Serial0/0/0

O IA 10.1.2.1 [110/193] via 34.1.1.1, 00:15:24, Serial0/0/0

O IA 10.1.3.1 [110/193] via 34.1.1.1, 00:15:24, Serial0/0/0

  12.0.0.0/30 is subnetted, 1 subnets

O IA 12.1.1.0 [110/192] via 34.1.1.1, 00:15:24, Serial0/0/0

  23.0.0.0/30 is subnetted, 1 subnets

O IA 23.1.1.0 [110/128] via 34.1.1.1, 00:15:24, Serial0/0/0

以上输出结果表明：nssa 区域可以接收区域间的路由。但在 Router2 上重发布的路由并没有出现在 Router4 的路由表上，这是为什么呢？

（5）查看 Router4 和 Router3 的链路状态数据库。

Router4#show   ip   ospf   database

    OSPF Router with ID (4.4.4.4) (Process ID 1)

    Router Link States (Area 2) //区域 2 类型 1 的 LSA

Link ID   ADV Router   Age   Seq#   Checksum Link count

| 4.4.4.4 | 4.4.4.4 | 1039 | 0x80000005 0x0089c3 2 |
| 3.3.3.3 | 3.3.3.3 | 1039 | 0x80000004 0x00ee63 2 |

Summary Net Link States (Area 2)   //区域 2 类型 3 的 LSA

Link ID	ADV Router	Age	Seq#	Checksum
23.1.1.0	3.3.3.3	1390	0x80000001	0x005d9f
12.1.1.0	3.3.3.3	1390	0x80000002	0x006c5a
10.1.1.1	3.3.3.3	1390	0x80000003	0x00962c
10.1.2.1	3.3.3.3	1390	0x80000004	0x008937
10.1.3.1	3.3.3.3	1390	0x80000005	0x007c42

Type-7 AS External Link States (Area 2)   //区域 2 类型 7 的 LSA

Link ID	ADV Router	Age	Seq#	Checksum Tag
192.168.1.0	4.4.4.4	1335	0x80000002	0x0070a4 0

//Router4 的链路状态数据库中没有类型 5 的 LSA

Router3#show ip ospf database

OSPF Router with ID (3.3.3.3) (Process ID 1)

Router Link States (Area 0)   //区域 0 类型 1 的 LSA

Link ID	ADV Router	Age	Seq#	Checksum Link count
3.3.3.3	3.3.3.3	1513	0x80000003	0x006909 2
2.2.2.2	2.2.2.2	600	0x80000003	0x00cfa5 2

Summary Net Link States (Area 0)   //区域 0 类型 3 的 LSA

Link ID	ADV Router	Age	Seq#	Checksum
34.1.1.0	3.3.3.3	1518	0x80000001	0x00cd24
12.1.1.0	2.2.2.2	1508	0x80000001	0x000a02
10.1.1.1	2.2.2.2	1508	0x80000002	0x0034d3
10.1.2.1	2.2.2.2	1508	0x80000003	0x0027de
10.1.3.1	2.2.2.2	1508	0x80000004	0x001ae9

Summary ASB Link States (Area 0)   //区域 0 类型 4 的 LSA

Link ID	ADV Router	Age	Seq#	Checksum
3.3.3.3	3.3.3.3	7	0x800000ec	0x0087be

Router Link States (Area 2)   //区域 2 类型 1 的 LSA

Link ID	ADV Router	Age	Seq#	Checksum	Link count
3.3.3.3	3.3.3.3	1153	0x80000004	0x00ee63	2
4.4.4.4	4.4.4.4	1153	0x80000005	0x0089c3	2

Summary Net Link States (Area 2) //区域 2 类型 3 的 LSA

Link ID	ADV Router	Age	Seq#	Checksum
23.1.1.0	3.3.3.3	1503	0x80000001	0x005d9f
12.1.1.0	3.3.3.3	1503	0x80000002	0x006c5a
10.1.1.1	3.3.3.3	1503	0x80000003	0x00962c
10.1.2.1	3.3.3.3	1503	0x80000004	0x008937
10.1.3.1	3.3.3.3	1503	0x80000005	0x007c42

Type-7 AS External Link States (Area 2) //区域 2 类型 7 的 LSA

Link ID	ADV Router	Age	Seq#	Checksum	Tag
192.168.1.0	4.4.4.4	1448	0x80000002	0x0070a4	0

Type-5 AS External Link States   //类型 5 的 LSA

Link ID	ADV Router	Age	Seq#	Checksum	Tag
100.1.1.0	2.2.2.2	600	0x80000001	0x005df4	0
192.168.1.0	4.4.4.4	1523	0x80000001	0x0094b1	0
192.168.1.0	3.3.3.3	1153	0x80000001	0x0043e2	0

Router3#

通过对比 Router3 和 Router4 上的链路状态数据库，我们得出这样的结论：

① nssa 区域允许接收 7 类 LSA 发送的外部路由；

② 区域边界路由器（ABR）没有能力将类型 5 的 LSA 转化成类型 7 的 LSA，却可以将类型 7 的 LSA 转变成类型 5 的 LSA。

如果禁止区域间的路由条目进入 nssa 区域，只允许区域边界路由器（ABR）注入一条默认路由，这样既可以减小路由表，节省带宽，同时也可以保证全网互通。其配置方法如下。

步骤 8　在 ABR 上添加 no-summary 选项。

（1）配置 Router3。

Router3#configure terminal

Router3(config)#router ospf 1

Router3(config-router)#area 2 nssa no-summary

Router3(config-router)#end

Router3#

（2）查看 Router4 路由。

Router4#show　ip　route

Codes: C - connected, S - static, I - IGRP, R - RIP, M - mobile, B - BGP

　　　　D - EIGRP, EX - EIGRP external, O - OSPF, IA - OSPF inter area

　　　　N1 - OSPF NSSA external type 1, N2 - OSPF NSSA external type 2

　　　　E1 - OSPF external type 1, E2 - OSPF external type 2, E - EGP

　　　　i - IS-IS, L1 - IS-IS level-1, L2 - IS-IS level-2, ia - IS-IS inter area

　　　　* - candidate default, U - per-user static route, o - ODR

　　　　P - periodic downloaded static route

Gateway of last resort is 34.1.1.1 to network 0.0.0.0

　　　34.0.0.0/30 is subnetted, 1 subnets

C　　　34.1.1.0 is directly connected, Serial0/0/0

C　　　192.168.1.0/24 is directly connected, Loopback1

O*IA 0.0.0.0/0 [110/65] via 34.1.1.1, 00:02:00, Serial0/0/0

OSPF 配置较为复杂，稍不留心就会出现各种故障。我们必须熟练使用各种排错命令。常用的有命令有以下几种。

① show ip protocols：查看已配置并运行的路由协议。

② show ip route：查看路由表。

③ show ip ospf interface：查看接口所属区域以及邻居。

④ show ip ospf neighbor：查看路由器的所有邻居。

⑤ debug ip ospf adj：查看 OSPF 路由器之间建立邻居关系的过程。

⑥ debug ip ospf events：查看 OSPF 事件。

⑦ debug ip ospf packet：查看 LSA 包的内容。

任务 4.5　路由重发布

【实训目的】

掌握 OSPF 和 RIP 之间的路由重发布，理解度量值 metric。

【实训任务】

1. 配置 OSPF。
2. 配置 RIP。
3. 配置 OSPF 和 RIP 之间的路由重发布。

【预备知识】

一、路由重发布的概念

在一个大型网络中,可能使用了多种路由协议,为了实现多种协议之间协同工作,路由器使用路由重分发(route redistribution)将从一种路由协议学习到的路由分发进另外一种路由协议。

路由重发布时要考虑管理距离和种子度量值。不同路由协议有不同的选路标准 metric,导致了不同路由协议之间不能互通。因此,路由重分发时要手工指定一个对方可以理解的 metric。

二、OSPF 和 RIP 之间的路由重发布命令

1. 将 OSPF 路由再发布进 RIP

Router(config)#router rip
Router(config-router)#redistribute   ospf   进程号   metric   {1-15}
#metric 随意指定,通常不要指定得太大以免后续 metric 超过 15

2. 将 RIP 路由再发布进 OSPF

Router(config)#router   ospf   进程号
Router(config-router)#redistribute   rip   subnets   [metric 数值][metric-type type-number]
#路由注入 ospf 时,默认 metric=20,类型为 OE2

【实训拓扑】

网络拓扑结构图如图 4-5 所示。

图 4-5　网络拓扑结构图

【实训设备】

路由器 3 台。

【实训步骤】

步骤 1　根据拓扑图搭建网络,并配置各路由器接口,参见表 4-5 所示。

表 4-5

设备名称	接口	IP	子网掩码
R1	S0/0/0	12.1.1.1	255.255.255.252
	Loopback 1	1.1.1.1	255.255.255.0
R2	S0/0/0	12.1.1.2	255.255.255.252
	S0/0/1	23.1.1.1	255.255.255.252
R3	S0/0/1	23.1.1.2	255.255.255.252
	Loopback 1	3.3.3.3	255.255.255.0

（1）配置 R1。

Router>enable

Router#configure terminal

Router(config)#hostname R1

R1(config)#interface Loopback 1

R1(config-if)#ip address 1.1.1.1 255.255.255.0

R1(config-if)#exit

R1(config)#interface s0/0/0

R1(config-if)#ip address 12.1.1.1 255.255.255.252

R1(config-if)#clock rate 64000

R1(config-if)#no shutdown

R1(config-if)#exit

R1(config)#

（2）配置 R2。

Router>enable

Router#configure terminal

Router(config)#hostname R2

R2(config)#interface s0/0/0

R2(config-if)#ip address 12.1.1.2 255.255.255.252

R2(config-if)#no shutdown

R2(config-if)#exit

R2(config)#interface s0/0/1

R2(config-if)#ip address 23.1.1.1 255.255.255.252

R2(config-if)#clock rate 64000

R2(config-if)#no shutdown

R2(config-if)#exit

R2(config)#

（3）配置 R3。

Router>enable

Router#configure terminal

Router(config)#hostname R3

R3(config)#interface s0/0/1

R3(config-if)#ip address 23.1.1.2 255.255.255.252

R3(config-if)#no shutdown

R3(config-if)#exit

R3(config)#interface loopback 1

R3(config-if)#ip address 3.3.3.3 255.255.255.0

R3(config-if)#exit

R3(config)#

步骤 2 配置路由协议。

（1）配置 R1。

R1(config)#router ospf 1

R1(config-router)#router-id 1.1.1.1

R1(config-router)#network 1.1.1.0 0.0.0.255 area 0

R1(config-router)#network 12.1.1.0 0.0.0.3 area 0

R1(config-router)#end

R1#

（2）配置 R2。

R2(config)#router ospf 1

R2(config-router)#router-id 2.2.2.2

R2(config-router)#network 12.1.1.0 0.0.0.3 area 0

R2(config-router)#exit

R2(config)#router rip

R2(config-router)#version 2

R2(config-router)#no auto-summary

R2(config-router)#network 23.0.0.0

R2(config-router)#exit

R2(config)#

（3）配置 R3。

R3(config)#router rip

R3(config-router)#version 2

R3(config-router)#no auto-summary

R3(config-router)#network   23.0.0.0

R3(config-router)#network   3.0.0.0

R3(config-router)#end

R3#

步骤 3　查看路由表。

（1）查看 R1 路由表。

R1#show   ip   route

Codes: C - connected, S - static, I - IGRP, R - RIP, M - mobile, B - BGP

　　　　D - EIGRP, EX - EIGRP external, O - OSPF, IA - OSPF inter area

　　　　N1 - OSPF NSSA external type 1, N2 - OSPF NSSA external type 2

　　　　E1 - OSPF external type 1, E2 - OSPF external type 2, E - EGP

　　　　i - IS-IS, L1 - IS-IS level-1, L2 - IS-IS level-2, ia - IS-IS inter area

　　　　* - candidate default, U - per-user static route, o - ODR

　　　　P - periodic downloaded static route

Gateway of last resort is not set

　　　　1.0.0.0/24 is subnetted, 1 subnets

C　　　1.1.1.0 is directly connected, Loopback1

　　　　12.0.0.0/30 is subnetted, 1 subnets

C　　　12.1.1.0 is directly connected, Serial0/0/0

以上输出结果表明 R1 没有学习到 RIP 路由。

（2）查看 R3 路由表。

R3#show   ip   route

Codes: C - connected, S - static, I - IGRP, R - RIP, M - mobile, B - BGP

　　　　D - EIGRP, EX - EIGRP external, O - OSPF, IA - OSPF inter area

　　　　N1 - OSPF NSSA external type 1, N2 - OSPF NSSA external type 2

　　　　E1 - OSPF external type 1, E2 - OSPF external type 2, E - EGP

　　　　i - IS-IS, L1 - IS-IS level-1, L2 - IS-IS level-2, ia - IS-IS inter area

　　　　* - candidate default, U - per-user static route, o - ODR

　　　　P - periodic downloaded static route

Gateway of last resort is not set

　　　　3.0.0.0/24 is subnetted, 1 subnets

C　　　3.3.3.0 is directly connected, Loopback1

　　　　23.0.0.0/30 is subnetted, 1 subnets

C　　　23.1.1.0 is directly connected, Serial0/0/1

以上输出结果表明 R3 没有学习到 OSPF 路由。

步骤 4　在 R2 上配置路由重发布。

R2(config)#router ospf 1

R2(config-router)#redistribute rip subnets //将 Rip 路由重发布进 OSPF

R2(config-router)#exit

R2(config)#router rip

R2(config-router)#redistribute ospf 1 metric 2 //将 OSPF 重发布进 RIP，metric 值默认为无限大，此处指定为 2

R2(config-router)#exit

R2(config)#

步骤 5 查看路由表。

（1）查看 R1 路由表。

R1#show ip route

Codes: C - connected, S - static, I - IGRP, R - RIP, M - mobile, B - BGP

      D - EIGRP, EX - EIGRP external, O - OSPF, IA - OSPF inter area

      N1 - OSPF NSSA external type 1, N2 - OSPF NSSA external type 2

      E1 - OSPF external type 1, E2 - OSPF external type 2, E - EGP

      i - IS-IS, L1 - IS-IS level-1, L2 - IS-IS level-2, ia - IS-IS inter area

      * - candidate default, U - per-user static route, o - ODR

      P - periodic downloaded static route

Gateway of last resort is not set

    1.0.0.0/24 is subnetted, 1 subnets

C   1.1.1.0 is directly connected, Loopback1

    3.0.0.0/24 is subnetted, 1 subnets

O E2   3.3.3.0 [110/20] via 12.1.1.2, 00:03:32, Serial0/0/0

    12.0.0.0/30 is subnetted, 1 subnets

C   12.1.1.0 is directly connected, Serial0/0/0

    23.0.0.0/30 is subnetted, 1 subnets

O E2   23.1.1.0 [110/20] via 12.1.1.2, 00:03:32, Serial0/0/0

以上输出结果表明 R1 学习到从 RIP 重发布来的两条路由，默认类型"O E2"，metric 值为 20。更改类型为"O E1"并指定 metric 值为 30 的命令为

R2(config-router)#redistribute rip subnets metric 30 metric-type 1

（2）查看 R3 路由表。

R3#show ip route

Codes: C - connected, S - static, I - IGRP, R - RIP, M - mobile, B - BGP

      D - EIGRP, EX - EIGRP external, O - OSPF, IA - OSPF inter area

      N1 - OSPF NSSA external type 1, N2 - OSPF NSSA external type 2

      E1 - OSPF external type 1, E2 - OSPF external type 2, E - EGP

i - IS-IS, L1 - IS-IS level-1, L2 - IS-IS level-2, ia - IS-IS inter area
* - candidate default, U - per-user static route, o - ODR
P - periodic downloaded static route

Gateway of last resort is not set

  1.0.0.0/32 is subnetted, 1 subnets
R   1.1.1.1 [120/2] via 23.1.1.1, 00:00:04, Serial0/0/1
  3.0.0.0/24 is subnetted, 1 subnets
C   3.3.3.0 is directly connected, Loopback1
  12.0.0.0/30 is subnetted, 1 subnets
R   12.1.1.0 [120/2] via 23.1.1.1, 00:00:04, Serial0/0/1
  23.0.0.0/30 is subnetted, 1 subnets
C   23.1.1.0 is directly connected, Serial0/0/1

以上输出结果表明 R3 学习到 OSPF 重发布进来的路由，metric 值均为手工指定的 2。

任务4.6 路由选择原则

【实训目的】

理解路由选择原则，合理配置路由。

【实训任务】

1. 配置 OSPF。
2. 配置 RIP。
3. 配置静态路由和默认路由。
4. 验证路由选择的原则。

【预备知识】

路由选择的原则按顺序如下：
1. 最长匹配优先；
2. 管理距离小者优先；
3. 度量值小者优先；
4. 默认路由。

【实训拓扑】

网络拓扑结构图如图 4-6-1 所示。

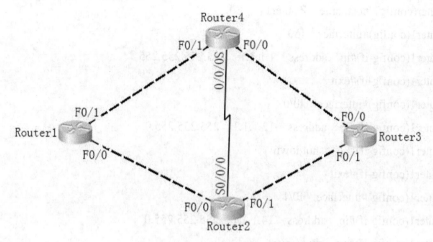

图 4-6-1 网络拓扑结构图

【实训设备】

路由器 4 台。

【实训步骤】

步骤 1 按照拓扑图构建网络，配置各路由器接口 IP 并开启，参见表 4-6 所示。

表 4-6

设备名称	接口	IP 地址	子网掩码
Router1	Loopback0	1.1.1.1	255.255.255.255
	F0/0	12.1.1.1	255.255.255.250
	F0/1	14.1.1.1	255.255.255.250
Router2	Loopback0	2.2.2.2	255.255.255.255
	F0/0	12.1.1.2	255.255.255.250
	F0/1	23.1.1.2	255.255.255.250
	S0/0/0	24.1.1.2	255.255.255.250
Router3	Loopback0	3.3.3.3	255.255.255.255
	F0/0	34.1.1.3	255.255.255.250
	F0/1	23.1.1.3	255.255.255.250
Router4	Loopback0	4.4.4.4	255.255.255.255
	F0/0	34.1.1.4	255.255.255.250
	F0/1	14.1.1.4	255.255.255.250
	S0/0/0	24.1.1.4	255.255.255.250

（1）配置路由器 Router1 接口 IP 并开启。

Router>enable

Router#configure terminal

Router(config)#hostname Router1

Router1(config)#interface lo0

Router1(config-if)#ip address 1.1.1.1 255.255.255.255

Router1(config-if)#exit

Router1(config)#interface f0/0

Router1(config-if)#ip address 12.1.1.1 255.255.255.0

Router1(config-if)#no shutdown

Router1(config-if)#exit

Router1(config)#interface f0/1

Router1(config-if)#ip address 14.1.1.1 255.255.255.0

Router1(config-if)#no shutdown

Router1(config-if)#exit

Router1(config)#

（2）配置路由器 Router2 接口 IP 并开启。

Router>enable

Router#configure terminal

Router(config)#hostname Router2

Router2(config)#interface lo0

Router2(config-if)#ip address 2.2.2.2 255.255.255.255

Router2(config-if)#exit

Router2(config)#interface f0/0

Router2(config-if)#ip address 12.1.1.2 255.255.255.0

Router2(config-if)#no shutdown

Router2(config-if)#exit

Router2(config)#interface f0/1

Router2(config-if)#ip address 23.1.1.2 255.255.255.0

Router2(config-if)#no shutdown

Router2(config-if)#

Router2(config-if)#exit

Router2(config)#interface s0/0/0

Router2(config-if)#ip address 24.1.1.2 255.255.255.0

Router2(config-if)#clock rate 128000

Router2(config-if)#no shutdown

Router2(config-if)#exit
Router2(config)#

(3) 配置路由器 Router3 接口 IP 并开启。

Router>enable
Router#configure terminal
Router(config)#hostname Router3
Router3(config)#interface f0/1
Router3(config-if)#ip address 23.1.1.3 255.255.255.0
Router3(config-if)#no shutdown
Router3(config-if)#exit
Router3(config)#interface lo0
Router3(config-if)#ip address 3.3.3.3 255.255.255.255
Router3(config-if)#exit
Router3(config)#interface f0/0
Router3(config-if)#ip address 34.1.1.3 255.255.255.0
Router3(config-if)#no shutdown
Router3(config-if)#exit
Router3(config)#

(4) 配置路由器 Router4 接口 IP 并开启。

Router>enable
Router#configure terminal
Router(config)#hostname Router4
Router4(config)#interface lo0
Router4(config-if)#ip address 4.4.4.4 255.255.255.255
Router4(config-if)#exit
Router4(config)#interface f0/0
Router4(config-if)#ip address 34.1.1.4 255.255.255.0
Router4(config-if)#no shutdown
Router4(config-if)#exit
Router4(config)#interface f0/1
Router4(config-if)#ip address 14.1.1.4 255.255.255.0
Router4(config-if)#no shutdown
Router4(config-if)#exit
Router4(config)#interface s0/0/0
Router4(config-if)#ip address 24.1.1.4 255.255.255.0
Router4(config-if)#no shutdown

Router4(config-if)#exit

Router4(config)#

步骤2 配置各路由器的路由表。

（1）配置路由器Router1。

Router1(config)#ip route 4.4.4.0 255.255.255.0 12.1.1.2

Router1(config)#ip route 0.0.0.0 0.0.0.0 14.1.1.4

（2）配置路由器Router2。

Router2(config)#ip route 0.0.0.0 0.0.0.0 12.1.1.1

Router2(config)#router ospf 1

Router2(config-router)#router-id 2.2.2.2

Router2(config-router)#network 23.1.1.0 0.0.0.255 area 0

Router2(config-router)#network 2.2.2.2 0.0.0.0 area 0

Router2(config-router)# default-information originate

Router2(config-router)#exit

Router2(config)#router rip

Router2(config-router)#version 2

Router2(config-router)#no auto-summary

Router2(config-router)#network 24.0.0.0

Router2(config-router)#end

Router2#

（3）配置路由器Router3。

Router3(config)#router ospf 1

Router3(config-router)#router-id 3.3.3.3

Router3(config-router)#network 23.1.1.0 0.0.0.255 area 0

Router3(config-router)#network 34.1.1.0 0.0.0.255 area 0

Router3(config-router)#network 3.3.3.3 0.0.0.0 area 0

Router3(config-router)#end

Router3#

（4）配置路由器Router4。

Router4(config)#router ospf 1

Router4(config-router)#router-id 4.4.4.4

Router4(config-router)#network 4.4.4.4 0.0.0.0 area 0

Router4(config-router)#network 34.1.1.0 0.0.0.255 area 0

Router4(config-router)#exit

Router4(config)#router rip

Router4(config-router)#version 2

Router4(config-router)#no   auto-summary

Router4(config-router)#network   24.0.0.0

Router4(config-router)# network   4.0.0.0 //在 RIP 中也通告环回口。

Router4(config-router)#end

Router4#

步骤 3 在 Router1 上做跟踪路由测试, 源 IP 为 1.1.1.1, 目的 IP 为 4.4.4.4。

Router1#traceroute

Protocol [ip]:

Target IP address: 4.4.4.4

Source address: 1.1.1.1

Numeric display [n]:

Timeout in seconds [3]:

Probe count [3]:

Minimum Time to Live [1]:

Maximum Time to Live [30]:

Type escape sequence to abort.

Tracing the route to 4.4.4.4

   1   *   31 msec   32 msec

   2   23.1.1.3   62 msec   62 msec   62 msec

   3   34.1.1.4   78 msec   94 msec   94 msec

经过观察发现, 源 IP 为 1.1.1.1 的数据包从路由器 Router1 的接口 F0/0 发送, 经过路由器 Router2 转发, 到达路由器 Router3, 最后到达 Router4, 发送给目的 IP:4.4.4.4。

想一想: 数据包为什么不从路由器 Router1 的接口 F0/1 发出直达路由器 Router4, 为何舍近求远呢? 要得到答案, 可以查看路由器 Router1 的路由表:

Router1# show   ip   route

Codes: C - connected, S - static, I - IGRP, R - RIP, M - mobile, B - BGP

       D - EIGRP, EX - EIGRP external, O - OSPF, IA - OSPF inter area

       N1 - OSPF NSSA external type 1, N2 - OSPF NSSA external type 2

       E1 - OSPF external type 1, E2 - OSPF external type 2, E - EGP

       i - IS-IS, L1 - IS-IS level-1, L2 - IS-IS level-2, ia - IS-IS inter area

       * - candidate default, U - per-user static route, o - ODR

       P - periodic downloaded static route

Gateway of last resort is 14.1.1.4 to network0.0.0.0

     1.0.0.0/32 is subnetted, 1 subnets

C   1.1.1.1 is directly connected, Loopback0

     4.0.0.0/24 is subnetted, 1 subnets

S   4.4.4.0 [1/0] via 12.1.1.2
         12.0.0.0/24 is subnetted, 1 subnets
C   12.1.1.0 is directly connected, FastEthernet0/0
         14.0.0.0/24 is subnetted, 1 subnets
C   14.1.1.0 is directly connected, FastEthernet0/1
S*   0.0.0.0/0 [1/0] via 14.1.1.4

经过仔细观察，路由器 Router1 存在一条静态路由 "S   4.4.4.0 [1/0] via 12.1.1.2"和一条默认路由 "S*   0.0.0.0/0 [1/0] via 14.1.1.4"。默认路由是找不到匹配的目的网络时的"无奈"选择，当存在到达目的网络的其他路由时，就不会考虑默认路由了。

在路由器 Router1 上增加一条路由指令：

Router1(config)#ip   route   4.4.4.4   255.255.255.255   14.1.1.4

再次查看路由器 Router1 的路由表：

Router1#show   ip   route
Codes: C - connected, S - static, I - IGRP, R - RIP, M - mobile, B - BGP
       D - EIGRP, EX - EIGRP external, O - OSPF, IA - OSPF inter area
       N1 - OSPF NSSA external type 1, N2 - OSPF NSSA external type 2
       E1 - OSPF external type 1, E2 - OSPF external type 2, E - EGP
       i - IS-IS, L1 - IS-IS level-1, L2 - IS-IS level-2, ia - IS-IS inter area
       * - candidate default, U - per-user static route, o - ODR
       P - periodic downloaded static route

Gateway of last resort is 14.1.1.4 to network 0.0.0.0

         1.0.0.0/32 is subnetted, 1 subnets
C   1.1.1.1 is directly connected, Loopback0
         4.0.0.0/8 is variably subnetted, 2 subnets, 2 masks
S   4.4.4.0/24 [1/0] via 12.1.1.2
S   4.4.4.4/32 [1/0] via 14.1.1.4
         12.0.0.0/24 is subnetted, 1 subnets
C   12.1.1.0 is directly connected, FastEthernet0/0
         14.0.0.0/24 is subnetted, 1 subnets
C   14.1.1.0 is directly connected, FastEthernet0/1
S*   0.0.0.0/0 [1/0] via 14.1.1.4

路由器 Router1 多出一条路由信息 "S   4.4.4.4/32 [1/0] via 14.1.1.4"。在模拟环境下测试，发现从源 IP：1.1.1.1 发出的数据包从路由器 Router1 的接口 F0/1 发出到达路由器 Router4，发送给目的 IP：4.4.4.4；应答数据包从 Router4 的接口 F0/0 发出，经过 Router3、Router2 转发到达 Router1，传送给 1.1.1.1。

路由选择的第一条原则：最长匹配优先。

步骤4 查看路由器 Router2 的路由表。

Router2#show ip route

Codes: C - connected, S - static, I - IGRP, R - RIP, M - mobile, B - BGP

   D - EIGRP, EX - EIGRP external, O - OSPF, IA - OSPF inter area

   N1 - OSPF NSSA external type 1, N2 - OSPF NSSA external type 2

   E1 - OSPF external type 1, E2 - OSPF external type 2, E - EGP

   i - IS-IS, L1 - IS-IS level-1, L2 - IS-IS level-2, ia - IS-IS inter area

   * - candidate default, U - per-user static route, o - ODR

   P - periodic downloaded static route

Gateway of last resort is 12.1.1.1 to network 0.0.0.0

  2.0.0.0/32 is subnetted, 1 subnets
C   2.2.2.2 is directly connected, Loopback0
  3.0.0.0/32 is subnetted, 1 subnets
O   3.3.3.3 [110/2] via 23.1.1.3, 01:22:10, FastEthernet0/1
  4.0.0.0/32 is subnetted, 1 subnets
O   4.4.4.4 [110/3] via 23.1.1.3, 01:10:18, FastEthernet0/1
  12.0.0.0/24 is subnetted, 1 subnets
C   12.1.1.0 is directly connected, FastEthernet0/0
  23.0.0.0/24 is subnetted, 1 subnets
C   23.1.1.0 is directly connected, FastEthernet0/1
  24.0.0.0/24 is subnetted, 1 subnets
C   24.1.1.0 is directly connected, Serial0/0/0
  34.0.0.0/24 is subnetted, 1 subnets
O   34.1.1.0 [110/2] via 23.1.1.3, 01:22:29, FastEthernet0/1
S* 0.0.0.0/0 [1/0] via 12.1.1.1

我们在前面配置路由器 Router4 时，故意在 RIP 下通告了环回口，而在路由器 Router2 上仅有一条 OSPF 路由 "O 4.4.4.4 [110/3] via 23.1.1.3, 01:10:18, FastEthernet0/1"，并没有显示出 RIP 路由。

路由选择的第二条原则：管理距离小者优先。

步骤5 在路由器 Router4 和 Router2 之间增加一条链路，配置接口 IP，并在 OSPF 中通告链路。如图 4-4-2 所示。

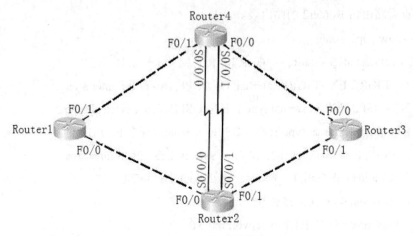

图 4-6-2 网络拓扑结构图

（1）路由器 Router4 配置接口 IP，更改 OSPF 的 COST 值，并在 OSPF 中通告链路。

Router4(config)#interface s0/0/1

Router4(config-if)#ip address 41.1.1.4 255.255.255.0

Router4(config-if)#clock rate 128000

Router4(config-if)# ip ospf cost 1

Router4(config-if)#no shutdown

Router4(config-if)#exit

Router4(config)#router ospf 1

Router4(config-router)#network 41.1.1.0 0.0.0.255 area 0

Router4(config-router)#end

（2）路由器 Router2 配置接口 IP,更改 OSPF 的 COST 值，并在 OSPF 中通告链路。

Router2(config)#interface s0/0/1

Router2(config-if)#ip address 41.1.1.2 255.255.255.0

Router2(config-if)#no shutdown

Router2(config-if)# ip ospf cost 1

Router2(config-if)#exit

Router2(config)#router ospf 1

Router2(config-router)#network 41.1.1.0 0.0.0.255 area 0

Router2(config-router)#end

（3）查看路由器 Router2 的路由表。

Router2#show ip route

Codes: C - connected, S - static, I - IGRP, R - RIP, M - mobile, B - BGP

      D - EIGRP, EX - EIGRP external, O - OSPF, IA - OSPF inter area

      N1 - OSPF NSSA external type 1, N2 - OSPF NSSA external type 2

      E1 - OSPF external type 1, E2 - OSPF external type 2, E - EGP

i - IS-IS, L1 - IS-IS level-1, L2 - IS-IS level-2, ia - IS-IS inter area

* - candidate default, U - per-user static route, o - ODR

P - periodic downloaded static route

Gateway of last resort is 12.1.1.1 to network0.0.0.0

  2.0.0.0/32 is subnetted, 1 subnets

C 2.2.2.2 is directly connected, Loopback0

  3.0.0.0/32 is subnetted, 1 subnets

O 3.3.3.3 [110/2] via 23.1.1.3, 01:51:33, FastEthernet0/1

  4.0.0.0/32 is subnetted, 1 subnets

O 4.4.4.4 [110/2] via 41.1.1.4, 00:02:17, Serial0/0/1

  12.0.0.0/24 is subnetted, 1 subnets

C 12.1.1.0 is directly connected, FastEthernet0/0

  23.0.0.0/24 is subnetted, 1 subnets

C 23.1.1.0 is directly connected, FastEthernet0/1

  24.0.0.0/24 is subnetted, 1 subnets

C 24.1.1.0 is directly connected, Serial0/0/0

  34.0.0.0/24 is subnetted, 1 subnets

O 34.1.1.0 [110/2] via 23.1.1.3, 01:51:52, FastEthernet0/1

    [110/2] via 41.1.1.4, 00:02:17, Serial0/0/1

  41.0.0.0/24 is subnetted, 1 subnets

C 41.1.1.0 is directly connected, Serial0/0/1

S* 0.0.0.0/0 [1/0] via 12.1.1.1

和前面步骤 4 中路由器 Router2 的路由信息相比，到路由器 Router4 环回口的路由信息变为"O 4.4.4.4 [110/2] via 41.1.1.4, 00:02:17, Serial0/0/1"。

路由选择的第三条原则：度量值小者优先。

当路由表中不存在明确的目的网段路由时，默认路由就成为最后的选择了。

项目 5

构建安全的网络

网络风险无处不在，机密数据的泄露、非授权用户私自登录网络设备更改配置等危险，无不对网络构成着重大的威胁。采取必要防范措施，提高网络安全系数成为网络管理员重中之重的工作。

通过本项目的实训，我们可以了解 PPP PAP 认证和 PPP CHAP 认证；理解 ACL 的执行过程，掌握标准 ACL、扩展 ACL 的配置方法。

任务 5.1 PPP PAP 认证

【实训目的】

配置 PPP PAP 认证。

【实训任务】

1. 配置 PPP PAP 单向认证。
2. 配置 PPP PAP 双向认证。

【预备知识】

一、PPP 概述

路由器是连接局域网和广域网的桥梁，无数的路由器构建了世界最大的网络——因特网。广域网链路的封装与以太网上的封装不同，PPP 是常见的广域网封装方式之一。

点到点协议（Point to Point Protocol，PPP）是 IETF（Internet Engineering Task Force，因特网工程任务组）推出的点到点类型线路的数据链路层协议，它支持多种网络层协议，支持认证、回拨、压缩、多链路捆绑等功能。

PAP（Password Authentication Protocol）密码验证协议的验证由被认证端发起，它向认证端发送用户名和密码；认证端查询自己的数据库，如果有匹配的用户名和密码，则验证通过，发送接收消息，否则拒绝连接。

二、PPP PAP 认证的配置

1. 配置认证方

Router(config)#username 用户名 password 密码

Router(config)#interface 串行接口

Router (config-if)#encapsulation ppp

Router (config-if)#ppp authentication pap

2. 配置被认证方

R2(config)#interface 串行接口

R2(config-if)#encapsulation ppp

R2(config-if)#ppp pap sent-username 用户名 password 密码

注意：认证方和被认证方双方的用户名和密码必须一致。

【实训拓扑】

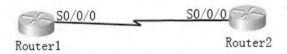

图 5-1 网络拓扑结构图

【实训设备】

路由器 2 台。

【实训步骤】

步骤 1 按网络拓扑结构图构建网络，配置路由器接口 IP 地址并开启，参见表 5-1 所示。

表 5-1

设备名称	接口	IP 地址	子网掩码
Router1	S0/0/0	12.1.1.1	255.255.255.252
Router2	S0/0/0	12.1.1.2	255.255.255.252

（1）配置路由器 Router1 接口 IP 地址并开启。

Router>enable

Router#configure terminal

Router(config)#hostname Router1

Router1(config)#interface s0/0/0

Router1(config-if)#ip address 12.1.1.1 255.255.255.252

Router1(config-if)#clock rate 128000 //Router1 是 DCE 端，需要配置时钟频率。

Router1(config-if)#no shutdown

Router1(config-if)#exit

Router1(config)#

（2）配置路由器 Router2 接口 IP 地址并开启。

Router>enable

Router#configure terminal

Router(config)#hostname Router2

Router2(config)#interface s0/0/0

Router2(config-if)#ip address 12.1.1.2 255.255.255.252

Router2(config-if)#no shutdown

Router2(config-if)#

步骤 2 配置 PPP PAP 单向认证，路由器 Router1 为认证方，路由器 Router2 为被认证方。

（1）配置路由器 Router1。

Router1(config)#username qkzz password xinxibu //配置用户名为 qkzz，密码为 xinxibu
Router1(config)#interface s0/0/0
Router1(config-if)#encapsulation ppp   //封装 ppp 协议
Router1(config-if)#ppp authentication pap //本路由器为认证方，认证方式为 pap 认证
Router1(config-if)#exit
Router1(config)#

（2）配置路由器 Router2。

Router2(config-if)#encapsulation ppp
Router2(config-if)#ppp pap sent-username qkzz password xinxibu //被认证方发送用户名为 qkzz，密码为 xinxibu
Router2(config-if)#end

（3）查看验证信息。

Router1#debug ppp authentication
PPP authentication debugging is on
Router1#configure terminal
Router1(config)#interface s0/0/0
Router1(config-if)#shutdown
Router1(config-if)#no shutdown
Router1(config-if)#
Serial0/0/0 PAP: I AUTH-REQ id 17 len 15
Serial0/0/0 PAP: Authenticating peer
Serial0/0/0 PAP: Phase is FORWARDING, Attempting Forward
%LINEPROTO-5-UPDOWN: Line protocol on Interface Serial0/0/0, changed state to up//验证通过，接口 s0/0/0 状态由 down 变为 up
Router1(config-if)#exit
Router1#debug ppp packet
PPP packet display debugging is on
Router1#
Serial0/0/0 PPP: I pkt type 0xc021, datagramsize 104
Serial0/0/0 PPP: O pkt type 0xc021, datagramsize 104
Serial0/0/0 PPP: I pkt type 0xc021, datagramsize 104
Serial0/0/0 PPP: I pkt type 0xc021, datagramsize 104
Serial0/0/0 PPP: O pkt type 0xc021, datagramsize 104

……

Router1#

Router1#no debug all   //关闭所有开启的 debug

All possible debugging has been turned off

Router1#

步骤 3 配置 PPP PAP 双向认证.。

在步骤 2 单向论证的基础上,我们再配置路由器 Router2 为认证方,路由器 Router1 为被认证方。

(1)配置路由器 Router2。

Router2>enable

Router2#configure terminal

Router2(config)#interface s0/0/0

Router2(config-if)#ppp authentication pap

%LINEPROTO-5-UPDOWN: Line protocol on Interface Serial0/0/0, changed state to down//由于被认证方尚未配置 pap 认证,接口 down 掉了

Router2(config)#username xinxibu password qkzz

Router2(config)#end

Router2#

(2)配置路由器 Router1。

Router1#configure terminal

Enter configuration commands, one per line. End with CNTL/Z.

Router1(config)#interface s0/0/0

Router1(config-if)#ppp pap sent-username xinxibu password qkzz

Router1(config-if)#

%LINEPROTO-5-UPDOWN: Line protocol on Interface Serial0/0/0, changed state to up//验证通过,接口 s0/0/0 状态由 down 变为 up

(3)验证连通性。

Router1#ping 12.1.1.1

Type escape sequence to abort.

Sending 5, 100-byte ICMP Echos to 12.1.1.1, timeout is 2 seconds:

!!!!!

Success rate is 100 percent (5/5), round-trip min/avg/max = 62/62/63 ms

Router1#

任务 5.2　配置 PPP CHAP 认证

【实训目的】

配置 PPP CHAP 认证。

【实训任务】

1. 配置 PPP CHAP 单向认证。
2. 配置 PPP CHAP 双向认证。

【预备知识】

一、CHAP 认证原理

CHAP（Challenge Handshake Authentication Protocol）质询握手协议，CHAP 不在线路上发送明文密码，而是发送经过摘要算法加工过的随机序列，被称为"挑战字符串"。此外，身份认证可以随时进行，包括在双方正常通信过程中。即使非法用户成功截获并破解了密码，此密码也将在一段时间内失效。因此，CHAP 认证比 PAP 认证更安全。

二、配置 CHAP 认证

1. 认证方配置命令

Router(config)#username　用户名　password　密码　//配置用户名、密码

Router(config)#interface　串行口

Router(config-if)#encapsulation　ppp　//封装 ppp 协议

Router(config-if)#ppp　authentication　chap　//本命令只在认证方配置，认证方式为 chap 认证

2. 被认证方配置命令

Router(config)#interface　串行口

Router(config-if)#encapsulation　ppp

Router(config-if)#ppp　chap　hostname　用户名　//配置被认证方用户名

Router(config-if)# ppp　chap　password　密码　//配置被认证方密码

三、简单配置 CHAP 双向认证

双方设备均要配置以下命令。

Router（config）#username　对方设备名 password 密码//双方认证密码完全一样

Router（config）#interface 串行口

Router（config）#encapsulation PPP //封装 ppp 协议

Router（config）#ppp authentication chap//PPP 认证方式为 chap 认证

【实训拓扑】

网络拓扑结构图如图 5-2 所示。

图 5-2　网络拓扑结构图

【实训设备】

路由器 2 台。

【实训步骤】

实训一

步骤 1　按网络拓扑结构图构建网络，配置路由器接口 IP 地址并开启，参见表 5-2-1 所示。

表 5-2-1

设备名称	接口	IP 地址	子网掩码
Router1	S0/0/0	12.1.1.1	255.255.255.252
Router2	S0/0/0	12.1.1.2	255.255.255.252

（1）配置路由器 Router1 接口 IP 地址并开启。

Router>enable

Router#configure terminal

Router(config)#hostname Router1

Router1(config)#interface s0/0/0

Router1(config-if)#ip address 12.1.1.1 255.255.255.252

Router1(config-if)#clock rate 128000 //Router1 是 DCE 端，需要配置时钟频率。

Router1(config-if)#no shutdown

Router1(config-if)#exit

Router1(config)#

（2）配置路由器 Router1 接口 IP 地址并开启。

Router>enable

Router#configure terminal

Router(config)#hostname Router2

Router2(config)#interface s0/0/0

Router2(config-if)#ip address 12.1.1.2 255.255.255.252

Router2(config-if)#no shutdown

Router2(config-if)#

步骤 2 配置 PPP CHAP 单向认证，路由器 Router1 为认证方，路由器 Router2 为被认证方。

（1）配置路由器 Router1。

Router1(config)#username qkzz password xinxibu //配置用户名为 qkzz，密码为 xinxibu

Router1(config)#interface s0/0/0

Router1(config-if)#encapsulation ppp //封装 ppp 协议

Router1(config-if)#ppp authentication chap //本命令只在认证方配置，认证方式为 chap 认证

Router1(config-if)#exit

Router1(config)#

（2）配置路由器 Router2。

Router2(config)#interface s0/0/0

Router2(config-if)#encapsulation ppp

Router2(config-if)#ppp chap hostname qkzz //被认证方用户名为 qkzz

Router2(config-if)# ppp chap password xinxibu //被认证方密码为 xinxibu

Router2(config-if)#end

Router2#

步骤 3 配置 PPP CHAP 双向认证。

在步骤 2 单向论证的基础上，再配置路由器 Router2 为认证方，路由器 Router1 为被认证方。

（1）配置路由器 Router2。

Router2>enable

Router2#configure terminal

Router2(config)#interface s0/0/0

Router2(config-if)#ppp authentication chap//本命令只在认证方配置，认证方式为 chap

Router2(config)#username xinxibu password qkzz

Router2(config)#end

Router2#

（2）配置路由器 Router1。

Router1#configure terminal

Enter configuration commands, one per line.　End with CNTL/Z.

Router1(config)#interface s0/0/0

Router1(config-if)#ppp　chap　hostname　xinxibu

Router1(config-if)#ppp　chap　password　qkzz

Router1(config-if)#

（3）验证连通性。

Router1#ping　12.1.1.1

Type escape sequence to abort.

Sending 5, 100-byte ICMP Echos to 12.1.1.1, timeout is 2 seconds:

!!!!!

Success rate is 100 percent (5/5), round-trip min/avg/max = 62/62/63 ms

Router1#

想一想：在 CHAP 双向认证中，假如双方设置认证的用户名均为对方设备的名称（hostname），并设置相同的密码，还需要在端口上应用命令"pppchap hostname"和"ppp chap password"吗？请读者按照后面实训二中的步骤自主进行验证。

实训二

步骤 1 按网络拓扑结构图构建网络，配置路由器接口 IP 地址并开启，参见表 5-2-2 所示。

表 5-2-2

设备名称	接口	IP 地址	子网掩码
Router1	S0/0/0	12.1.1.1	255.255.255.252
Router2	S0/0/0	12.1.1.2	255.255.255.252

（1）配置路由器 Router1 接口 IP 地址并开启。

Router>enable

Router#configure　terminal

Router(config)#hostname　Router1

Router1(config)#interface　s0/0/0

Router1(config-if)#ip　address　12.1.1.1　255.255.255.252

Router1(config-if)#clock　rate　128000　　//Router1 是 DCE 端，需要配置时钟频率

Router1(config-if)#no　shutdown

Router1(config-if)#exit

Router1(config)#

（2）配置路由器 Router1 接口 IP 地址并开启。

Router>enable

Router#configure　terminal

Router(config)#hostname　Router2

Router2(config)#interface　s0/0/0

Router2(config-if)#ip address 12.1.1.2 255.255.255.252

Router2(config-if)#no shutdown

Router2(config-if)#

步骤 2 配置 PPP CHAP 双向认证。

（1）配置路由器 Router1。

Router1(config)#username Router2 password xinxibu //配置用户名为 Router 2，密码为 xinxibu

Router1(config)#interface s0/0/0

Router1(config-if)#encapsulation ppp　　//封装 ppp 协议

Router1(config-if)#ppp authentication chap //配置认证方式为 chap 认证

Router1(config-if)#end

Router1#

（2）配置路由器 Router2。

Router2(config)#username Router1 password xinxibu //配置用户名为 Router1，密码为 xinxibu

Router2(config)#interface s0/0/0

Router2(config-if)#encapsulation ppp

Router2(config-if)# ppp authentication chap //配置认证方式为 chap 认证

Router2(config-if)#end

Router2#

步骤 3 验证连通性。

Router1#ping 12.1.1.1

Type escape sequence to abort.

Sending 5, 100-byte ICMP Echos to 12.1.1.1, timeout is 2 seconds:

!!!!!

Success rate is 100 percent (5/5), round-trip min/avg/max = 62/62/63 ms

Router1#

以上是配置 CHAP 双向认证最简单的方式，要求配置用户名必须为对方路由器的主机名，认证方和被认证方的密码也必须保持一致。

任务 5.3　标准 ACL

【实训目的】

配置标准 ACL，过滤掉不符合条件的数据包，理解基于 IP 源地址的包过滤原理和应用方法，掌握标准访问控制列表的配置。

【实训任务】

1. 根据拓扑图连接网络设备，构建局域网。
2. 配置交换机和路由器的路由，实现网络互通。
3. 配置标准 ACL，并应用到路由器的接口上。
4. 验证测试配置标准 ACL 后的效果。

【预备知识】

一、标准访问控制列表作用

标准 ACL 只检查数据包的源地址。网络管理员可以使用标准 ACL 阻止或允许来自某一网络的所有通信流量，还可以拒绝某一协议簇（如 IP）的所有通信流量。

注意：访问控制列表末尾隐含一条拒绝所有流量的指令，所以，访问控制列表至少要有一条允许数据通过的命令语句。

二、标准访问控制列表的基本语法

标准访问控制列表语法：
Access-list　访问控制列表号　{deny | permit}　源网络地址　[通配符掩码]
access-list 命令参数的含义如下。

（1）标准 ACL 的号码范围是 1~99。

（2）是必选项，deny 表示如果满足条件，则数据包被丢弃；permit 表示如果满足条件，则数据包被允许通过该接口。

（3）通配符掩码（反掩码），可选项。在通配符掩码位中，0 表示"检查相应的位"，1 表示"不检查相应位"。子网掩码按位求反就可以得到相应的通配符掩码。

如：十进制子网掩码 255.255.255.0 转化成二进制时为

11111111　11111111　11111111　00000000

将每一位的数值求反，则变为

00000000　00000000　00000000　11111111

将上面二进制转化成十进制则通配符掩码为

0.0.0.255

我们可以在"access-list"命令前加"no"，用来删除一个已经建立的标准 ACL。

三、关键字 any 和 host 的用法

（1）any：允许源地址为任意 IP 地址的数据包通过。

access-list　1　permit　any 等价于下面的命令：

access-list 1 permit 0.0.0.0 255.255.255.255

（2）host：仅允许单台主机的流量通过。如以下命令：

access-list 1 permit host 192.168.100.100

四、在接口上应用 ACL

语法：ip access-group 访问控制列表号 {in/out}

（1）in：数据包从外部进入设备。

（2）out：数据包从设备流向外部。

五、ACL 的执行过程

ACL 是一组控制命令的列表，它从第一条命令开始检查，如果匹配规定的条件，则终止比较后面的控制命令，转而执行该命令规定的操作，允许或阻止符合条件的数据包通过接口。如果第一条命令不符合规定的条件，则继续检查第二条命令……重复以上过程，直到最后 ACL 有一条默认的隐含命令，阻止所有数据通过。

ACL 规则助记词：

从顶向下，

匹配即停；

允许/拒绝，

按章执行。

【实训拓扑】

网络拓扑结构图如图 5-3 所示。

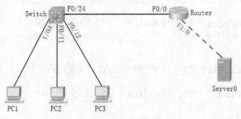

图 5-3 网络拓扑结构图

【实训设备】

三层交换机 1 台、路由器 1 台、计算机 4 台。

【实训步骤】

步骤 1 配置各设备的接口 IP，参见表 5-3 所示。

表 5-3

设备名称	接口	IP	子网掩码
Switch	Vlan10(f0/1-10)	192.168.10.1	255.255.255.0
	Vlan20(f0/11-20)	192.168.20.1	255.255.255.0
	F0/24	192.168.30.1	255.255.255.0
Router	F0/0	192.168.30.2	255.255.255.0
	F1/0	200.1.1.1	255.255.255.0
PC1	网卡	192.168.10.2	255.255.255.0
PC2	网卡	192.168.20.2	255.255.255.0
Server0	网卡	200.1.1.2	255.255.255.0
PC3	网卡	192.168.20.3	255.255.255.0

（1）配置交换机的接口 IP。

Switch>enable

Switch#configure terminal

Switch(config)#vlan 10

Switch(config-vlan)#vlan 20

Switch(config-vlan)#exit

Switch(config)#interface range f0/1-10

Switch(config-if-range)#switchport access vlan 10

Switch(config-if-range)#exit

Switch(config)#interface range f0/11-20

Switch(config-if-range)#switchport access vlan 20

Switch(config-if-range)#exit

Switch(config)#interface vlan 10   //配置 VLAN10 的 SVI 接口 IP

Switch(config-if)#ip address 192.168.10.1 255.255.255.0

Switch(config-if)#exit

Switch(config)#interface vlan 20   //配置 VLAN20 的 SVI 接口 IP

Switch(config-if)#ip address 192.168.20.1 255.255.255.0

Switch(config-if)#exit

Switch(config)#interface f0/24

Switch(config-if)#no switchport   //ACCESS 模式转变为路由模式

Switch(config-if)#ip address 192.168.30.1 255.255.255.0

Switch(config-if)#exit

Switch(config)#

（2）配置路由器的接口 IP。

Router>enable

Router#configure Terminal

Router(config)#interface   f0/0

Router(config-if)#ip   address   192.168.30.2   255.255.255.0

Router(config-if)#no   shutdown

Router(config-if)#exit

Router(config)#interface   f1/0

Router(config-if)#ip   address   200.1.1.1   255.255.255.0

Router(config-if)#no   shutdown

Router(config-if)#exit

Router(config)#

（3）验证测试 PC1 和 PC2 的连通性。

PC>ping   192.168.20.2

Pinging 192.168.20.2 with 32 bytes of data:

Reply from 192.168.20.2: bytes=32 time=63ms TTL=127

Reply from 192.168.20.2: bytes=32 time=63ms TTL=127

Reply from 192.168.20.2: bytes=32 time=63ms TTL=127

Reply from 192.168.20.2: bytes=32 time=62ms TTL=127

Ping statistics for 192.168.20.2:

　　Packets: Sent = 4, Received = 4, Lost = 0 (0% loss),

Approximate round trip times in milli-seconds:

　　Minimum = 62ms, Maximum = 63ms, Average = 62ms

以上结果表明，PC1 能够 ping 通 PC2；同样，PC1 也能够 ping 通 PC3。接下来测试 PC1 和 server0 的连通性：

PC>ping   200.1.1.2

Pinging 200.1.1.2 with 32 bytes of data:

Reply from 192.168.10.1: Destination host unreachable.

Reply from 192.168.10.1: Destination host unreachable.

Reply from 192.168.10.1: Destination host unreachable.

Reply from 192.168.10.1: Destination host unreachable.

Ping statistics for 200.1.1.2:

　　Packets: Sent = 4, Received = 0, Lost = 4 (100% loss)

结果显示，计算机 PC1 无法和 server0 通信，请同学们思考其中的原因。

步骤 2 配置各设备的路由.

（1）配置交换机的路由。

Switch(config)#ip   routing

Switch(config)#ip   route   0.0.0.0 0.0.0.0   192.168.30.2

（2）配置路由器的路由。

Router(config)#ip route   192.168.10.0   255.255.255.0   192.168.30.1
Router(config)#ip route   192.168.20.0   255.255.255.0   192.168.30.1

（3）观察交换机和路由器的路由表。

Switch#show ip route

Codes: C - connected, S - static, I - IGRP, R - RIP, M - mobile, B - BGP

   D - EIGRP, EX - EIGRP external, O - OSPF, IA - OSPF inter area

   N1 - OSPF NSSA external type 1, N2 - OSPF NSSA external type 2

   E1 - OSPF external type 1, E2 - OSPF external type 2, E - EGP

   i - IS-IS, L1 - IS-IS level-1, L2 - IS-IS level-2, ia - IS-IS inter area

   * - candidate default, U - per-user static route, o - ODR

   P - periodic downloaded static route

Gateway of last resort is 192.168.30.2 to network0.0.0.0

C 192.168.10.0/24 is directly connected, Vlan10
C 192.168.20.0/24 is directly connected, Vlan20
C 192.168.30.0/24 is directly connected, FastEthernet0/24
S* 0.0.0.0/0 [1/0] via 192.168.30.2

以上结果显示交换机有三条直连路由（C）:192.168.10.0、192.168.20.0、192.168.30.0。此外还有一条默认路由（S*）:0.0.0.0，它的下一跳地址是 192.168.30.2。

Router#show ip route

Codes: C - connected, S - static, I - IGRP, R - RIP, M - mobile, B - BGP

   D - EIGRP, EX - EIGRP external, O - OSPF, IA - OSPF inter area

   N1 - OSPF NSSA external type 1, N2 - OSPF NSSA external type 2

   E1 - OSPF external type 1, E2 - OSPF external type 2, E - EGP

   i - IS-IS, L1 - IS-IS level-1, L2 - IS-IS level-2, ia - IS-IS inter area

   * - candidate default, U - per-user static route, o - ODR

   P - periodic downloaded static route

Gateway of last resort is not set

S 192.168.10.0/24 [1/0] via 192.168.30.1
S 192.168.20.0/24 [1/0] via 192.168.30.1
C 192.168.30.0/24 is directly connected, FastEthernet0/0
C 200.1.1.0/24 is directly connected, FastEthernet1/0

结果显示路由器有两条直连路由（C）和两条静态路由（S）。

想一想：两条静态路由的下一跳地址分别是什么？静态路由的管理距离是多少？

（4）请同学们自己验证测试 pc1、PC2、PC3 和 PC4 的连通性。

步骤 3 在路由器上配置标准 ACL，只允许源 IP 地址为 192.168.10.0/24 的所有计算机和 192.168.20.2 的计算机能够访问计算机 server0。

Router(config)#access-list 1 permit 192.168.10.0 0.0.0.255　//允许源网络为 192.168.10.0 的所有主机

Router(config)#access-list 1 permit host 192.168.20.2　//允许主机 192.168.20.2

Router(config)#access-list 1 deny any　//拒绝所有 IP，这是 ACL 的最后一条默认隐含命令

Router(config)#interface f1/0

Router(config-if)#ip access-group 1 out　//在接口 f1/0 的出方向应用 ACL

步骤 4　分别测试 PC1、PC2、PC3 与 SERVER0 的连通性。

（1）测试 PC1 与 SERVER0 的连通性：

PC>ping　200.1.1.2

Pinging 200.1.1.2 with 32 bytes of data:

Reply from 200.1.1.2: bytes=32 time=79ms TTL=126

Reply from 200.1.1.2: bytes=32 time=94ms TTL=126

Reply from 200.1.1.2: bytes=32 time=80ms TTL=126

Reply from 200.1.1.2: bytes=32 time=78ms TTL=126

Ping statistics for 200.1.1.2:

　　Packets: Sent = 4, Received = 4, Lost = 0 (0% loss),

Approximate round trip times in milli-seconds:

Minimum = 78ms, Maximum = 94ms, Average = 82ms

（2）测试 PC2 与 SERVER0 的连通性：

PC>ping 200.1.1.2

Pinging 200.1.1.2 with 32 bytes of data:

Reply from 200.1.1.2: bytes=32 time=63ms TTL=126

Reply from 200.1.1.2: bytes=32 time=78ms TTL=126

Reply from 200.1.1.2: bytes=32 time=94ms TTL=126

Reply from 200.1.1.2: bytes=32 time=78ms TTL=126

Ping statistics for 200.1.1.2:

　　Packets: Sent = 4, Received = 4, Lost = 0 (0% loss),

Approximate round trip times in milli-seconds:

Minimum = 63ms, Maximum = 94ms, Average = 78ms

（3）测试 PC3 与 SERVER0 的连通性：

PC>ping　200.1.1.2

Pinging 200.1.1.2 with 32 bytes of data:

Reply from 192.168.30.2: Destination host unreachable.

Reply from 192.168.30.2: Destination host unreachable.

Reply from 192.168.30.2: Destination host unreachable.

Reply from 192.168.30.2: Destination host unreachable.
Ping statistics for 200.1.1.2:

　　Packets: Sent = 4, Received = 0, Lost = 4 (100% loss)

以上输出结果显示 PC1、PC2 和 SERVER0 能够互通，而 PC3 的数据包到达路由器，经 f1/0 向外转发时因为不符合 ACL 设置的转发条件而被丢弃，故而 PC3 无法和 SERVER0 通信。

任务5.4　扩展 ACL

【实训目的】

通过配置扩展 ACL，理解基于 IP 地址、协议和端口的包过滤原理和应用方法，使学生掌握扩展访问控制列表的配置方法。

【实训任务】

1. 根据拓扑图连接网络设备，构建局域网。
2. 配置扩展 ACL，实现数据包的过滤。
3. 配置标准 ACL，理解和扩展 ACL 的区别。
4. 测试网络连通性。

【预备知识】

一、扩展访问控制列表的作用

和标准 ACL 相比，扩展 ACL 既检查数据包的源地址，也检查数据包的目的地址，同时还可以检查数据包的特定协议类型、端口号等。

扩展 ACL 比标准 ACL 提供了更广泛的控制范围。例如，网络管理员如果希望做到"允许外来的 Web 通信流量通过，拒绝外来的 FTP 和 Telnetwork 等通信流量"，那么，他可以使用扩展 ACL 来达到目的，因为标准 ACL 不能控制得这么精确。

二、扩展访问控制列表的语法

扩展访问控制列表语法：

access-list 访问控制列表号 {deny|permit} 协议类型　源网络地址　[通配符掩码]　目的网络地址 [通配符掩码]　[运算符　端口号]

命令参数的含义如下。

（1）扩展访问控制列表号范围为 100～199。

（2）{deny/permit}为必选项，在 deny 和 permit 中二选其一。deny 表示如果满足条件，则数据包将被丢弃；permit 表示如果满足条件，则数据包被允许通过。

（3）协议类型包括 IP、TCP、UDP、ICMP、OSPF 等。

（4）运算符有 lt、gt、eq、neq，分别表示"小于、大于、等于、不等于"。

（5）端口号：一是物理端口，比如交换机、路由器用于连接其他网络设备的接口，如 RJ-45 端口、serial 端口等。二是逻辑意义上的端口，一般是指 TCP/IP 协议中的端口，端口号的范围为 0 ~ 65535，比如用于浏览网页服务的 80 端口，用于 FTP 服务的 21 端口等。

三、常用协议及协议的端口号

扩展访问控制列表相对于基本访问控制列表有哪些不同呢？我们可以使用扩展 ACL 来做针对源地址、目的地址、协议及端口号的包过滤。

常见的端口号及其对应协议见表 5-4-1 所示。

表 5-4-1

协议	端口号
Telnet	23
FTP	20/21
SMTP	25
POP3	110
DNS	53
Http	80
TFTP	69

【实训拓扑】

网络拓扑结构图如图 5-4 所示。

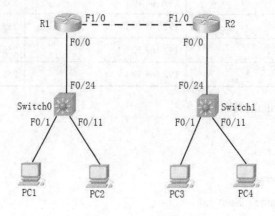

图 5-4　网络拓扑结构图

【实训设备】

路由器 2 台、三层交换机 2 台、计算机 4 台。

【实训步骤】

步骤 1 配置各设备的接口 IP 地址，参见表 5-4-2 所示。

表 5-4-2

设备名称	接口	IP	子网掩码
R1	F0/0	172.16.1.2	255.255.255.0
	F1/0	12.1.1.1	255.255.255.252
R2	F0/0	172.16.2.2	255.255.255.0
	F1/0	12.1.1.2	255.255.255.252
Switch0	Vlan10 (F0/1-10)	192.168.10.254	255.255.255.0
	Vlan20 (F0/11-20)	192.168.20.254	255.255.255.0
	F0/24	172.16.1.1	255.255.255.0
Switch1	Vlan30 (F0/1-10)	192.168.30.254	255.255.255.0
	Vlan40 (F0/11-20)	192.168.40.254	255.255.255.0
	F0/24	172.16.2.1	255.255.255.0
PC1	网卡	192.168.10.1	255.255.255.0
PC2	网卡	192.168.20.1	255.255.255.0
PC3	网卡	192.168.30.1	255.255.255.0
PC4	网卡	192.168.40.1	255.255.255.0

（1）配置交换机 swith0 的接口 IP。

Switch>enable

Switch#

Switch#configure   terminal

Switch(config)#hostname   Switch0

Switch0(config)#vlan   10

Switch0(config-vlan)#vlan   20

Switch0(config-vlan)#exit
Switch0(config)#interface vlan 10
Switch0(config-if)#ip address 192.168.10.254 255.255.255.0
Switch0(config-if)#exit
Switch0(config)#interface vlan 20
Switch0(config-if)#ip address 192.168.20.254 255.255.255.0
Switch0(config-if)#exit
Switch0(config)#interface f0/24
Switch0(config-if)#no switchport
Switch0(config-if)#ip address 172.16.1.1 255.255.255.0
Switch0(config-if)#exit
Switch0(config)#
Switch0(config)#interface range f0/1-10
Switch0(config-if-range)#switchport access vlan 10
Switch0(config-if-range)#exit
Switch0(config)#interface range f0/11-20
Switch0(config-if-range)#switchport access vlan 20
Switch0(config-if-range)#exit
Switch0(config)#

（2）配置路由器 R1 的接口 IP。

Router>enable
Router#configure terminal
Router(config)#hostname R1
R1(config)#interface f0/0
R1(config-if)#ip address 172.16.1.2 255.255.255.0
R1(config-if)#no shutdown
R1(config-if)#exit
R1(config)#interface f1/0
R1(config-if)#ip address 12.1.1.1 255.255.255.252
R1(config-if)#no shutdown
R1(config-if)#exit
R1(config)#

（3）配置路由器 R2 的接口 IP。

Router>enable
Router#configure terminal
Router(config)#hostname R2

R2(config)#interface   f0/0

R2(config-if)#ip   address   172.16.2.2   255.255.255.0

R2(config-if)#no   shutdown

R2(config-if)#exit

R2(config)#interface   f1/0

R2(config-if)#ip   address   12.1.1.2   255.255.255.252

R2(config-if)#no   shutdown

R2(config-if)#exit

R2(config)#

（4）配置交换机 swith0 的接口 IP。

Switch>enable

Switch#configure   terminal

Switch(config)#hostname   Switch1

Switch1(config)#vlan   30

Switch1(config-vlan)#vlan   40

Switch1(config-vlan)#exit

Switch1(config)#interface   vlan 30

Switch1(config-if)#ip   address   192.168.30.254   255.255.255.0

Switch1(config-if)#exit

Switch1(config)#interface   vlan 40

Switch1(config-if)#ip   address   192.168.40.254   255.255.255.0

Switch1(config-if)#exit

Switch1(config)#interface   f0/24

Switch1(config-if)#no   switchport

Switch1(config-if)#ip   address   172.16.2.1   255.255.255.0

Switch1(config-if)#exit

Switch1(config)#interface   range   f0/1-10

Switch1(config-if-range)#switchport   access   vlan 30

Switch1(config-if-range)#exit

Switch1(config)#interface   range   f0/11-20

Switch1(config-if-range)#switchport   access   vlan 40

Switch1(config-if-range)#exit

Switch1(config)#

步骤 2 配置各网络设备的路由协议。

（1）配置交换机 switch0 的路由协议 OSPF。

Switch0(config)#router   ospf   1

Switch0(config-router)#router-id   1.1.1.1
Switch0(config-router)#network   192.168.10.1 0.0.0.255   area   1
Switch0(config-router)#network   192.168.20.1 0.0.0.255   area   1
Switch0(config-router)#network   172.16.1.0   0.0.0.255   area   1
Switch0(config-router)#

（2）配置路由器 R1 的路由协议 OSPF。

R1(config)#router   ospf   1
R1(config-router)#router-id   2.2.2.2
R1(config-router)#network   172.16.1.0 0.0.0.255   area   1
R1(config-router)#network   12.1.1.0 0.0.0.3   area   0
R1(config-router)#

（3）配置路由器 R2 的路由协议 OSPF。

R2(config)#router   ospf   1
R2(config-router)#router-id 3.3.3.3
R2(config-router)#network   12.1.1.0   0.0.0.3   area   0
R2(config-router)#network   172.16.2.0 0.0.0.255   area   2
R2(config-router)#end

（4）配置交换机 Switch1 的路由协议 OSPF。

Switch(config)#router   ospf   1
Switch(config-router)#router-id   4.4.4.4
Switch(config-router)#network   192.168.30.0   0.0.0.255   area   2
Switch(config-router)#network   192.168.40.0   0.0.0.255   area   2
Switch(config-router)#network   172.16.2.0   0.0.0.255   area   2
Switch(config-router)#end

（5）查看交换机 Swith0 的路由表。

Switch0#show   ip   route
Codes: C - connected, S - static, I - IGRP, R - RIP, M - mobile, B - BGP
       D - EIGRP, EX - EIGRP external, O - OSPF, IA - OSPF inter area
       N1 - OSPF NSSA external type 1, N2 - OSPF NSSA external type 2
       E1 - OSPF external type 1, E2 - OSPF external type 2, E - EGP
       i - IS-IS, L1 - IS-IS level-1, L2 - IS-IS level-2, ia - IS-IS inter area
       * - candidate default, U - per-user static route, o - ODR
       P - periodic downloaded static route

Gateway of last resort is not set

     12.0.0.0/30 is subnetted, 1 subnets
O IA   12.1.1.0 [110/2] via 172.16.1.2, 00:05:04, FastEthernet0/24

172.16.0.0/24 is subnetted, 2 subnets

C 172.16.1.0 is directly connected, FastEthernet0/24

O IA 172.16.2.0 [110/3] via 172.16.1.2, 00:05:04, FastEthernet0/24

C 192.168.10.0/24 is directly connected, Vlan10

C 192.168.20.0/24 is directly connected, Vlan20

O IA 192.168.30.0/24 [110/4] via 172.16.1.2, 00:00:55, FastEthernet0/24

O IA 192.168.40.0/24 [110/4] via 172.16.1.2, 00:03:22, FastEthernet0/24

Switch#

请同学们验证测试四台计算机间能否互通。如果上述配置正确，则此时四台计算机可以互相通信。

步骤 3　利用扩展 ACL 禁止 vlan20 和 vlan 30 互访，其他不受限制。

（1）在交换机 Swith0 上配置扩展 ACL。

Switch0(config)#access-list 100 deny ip 192.168.20.0 0.0.0.255 192.168.30.0 0.0.0.255

Switch0(config)#access-list 100 permit ip any any

Switch0(config)#interface vlan 20

Switch(config-if)#ip access-group 100 in

（2）在交换机 Swith1 上配置扩展 ACL。

Switch1(config)#access-list 100 deny ip 192.168.30.0 0.0.0.255 192.168.20.0 0.0.0.255

Switch1(config)#access-list 100 permit ip any any

Switch1(config)#interface vlan 30

Switch1(config-if)#ip access-group 100 in

Switch1(config-if)#

步骤 4　测试各计算机连通性。

（1）测试 PC2 和 PC3 的连通性：

PC>ping 192.168.30.1

Pinging 200.1.1.2 with 32 bytes of data:

Reply from 192.168.20.1: Destination host unreachable.

Reply from 192.168.20.1: Destination host unreachable.

Reply from 192.168.20.1: Destination host unreachable.

Reply from 192.168.20.1: Destination host unreachable.

Ping statistics for 192.168.30.1:

  Packets: Sent = 4, Received = 0, Lost = 4 (100% loss)

（2）测试 PC2 和 PC1 的连通性：

PC>ping 192.168.10.1

Pinging 192.168.10.1 with 32 bytes of data:

Reply from 192.168.10.1: bytes=32 time=63ms TTL=126

Reply from 192.168.10.1: bytes=32 time=78ms TTL=126

Reply from 192.168.10.1 bytes=32 time=94ms TTL=126

Reply from 192.168.10.1: bytes=32 time=78ms TTL=126

Ping statistics for 192.168.10.1:

    Packets: Sent = 4, Received = 4, Lost = 0 (0% loss),

Approximate round trip times in milli-seconds:

Minimum = 63ms, Maximum = 94ms, Average = 78ms

（3）测试 PC2 和 PC4 的连通性：

PC>ping 192.168.40.1

Pinging 192.168.40.1 with 32 bytes of data:

Reply from 192.168.40.1: bytes=32 time=63ms TTL=126

Reply from 192.168.40.1: bytes=32 time=78ms TTL=126

Reply from 192.168.40.1 bytes=32 time=94ms TTL=126

Reply from 192.168.40.1: bytes=32 time=78ms TTL=126

Ping statistics for 192.168.40.1:

    Packets: Sent = 4, Received = 4, Lost = 0 (0% loss),

Approximate round trip times in milli-seconds:

Minimum = 63ms, Maximum = 94ms, Average = 78ms

经过以上测试可以看出，配置扩展访问列表后，PC2 无法和 pc3 通信，而和其他计算机间的通信正常。

步骤 5　本实训效果也可以使用标准 ACL 实现。

配置命令如下。

（1）交换机 switch0。

Switch0(config)#access-list　1　deny　192.168.30.0 0.0.0.255

Switch0(config)#access-list　1　permit　any

Switch0(config)#interface　vlan　20

Switch0(config-if)#no　ip　access-group　100　in　//撤销应用扩展 ACL 100

Switch0(config-if)#ip　access-group　1　out

（2）交换机 switch1。

Switch1(config)#access-list　1　deny　192.168.20.0 0.0.0.255

Switch1(config)#access-list 1　permit　any

Switch1(config)#interface　vlan　30

Switch0(config-if)#no　ip　access-group　100　in　//撤销应用扩展 ACL 100

Switch1(config-if)#ip　access-group　1　out

（3）测试 PC2 与 PC1、PC3、PC4 间的连通性

PC>ping 192.168.10.1

Pinging 192.168.10.1 with 32 bytes of data:

Reply from 192.168.10.1: bytes=32 time=63ms TTL=126

Reply from 192.168.10.1: bytes=32 time=78ms TTL=126

Reply from 192.168.10.1 bytes=32 time=94ms TTL=126

Reply from 192.168.10.1: bytes=32 time=78ms TTL=126

Ping statistics for 192.168.10.1:

    Packets: Sent = 4, Received = 4, Lost = 0 (0% loss),

Approximate round trip times in milli-seconds:

Minimum = 63ms, Maximum = 94ms, Average = 78ms

PC>ping   192.168.30.1

Pinging 200.1.1.2 with 32 bytes of data:

Reply from 192.168.20.1: Destination host unreachable.

Reply from 192.168.20.1: Destination host unreachable.

Reply from 192.168.20.1: Destination host unreachable.

Reply from 192.168.20.1: Destination host unreachable.

Ping statistics for 192.168.30.1:

    Packets: Sent = 4, Received = 0, Lost = 4 (100% loss)

PC>ping 192.168.40.1

Pinging 192.168.40.1 with 32 bytes of data:

Reply from 192.168.40.1: bytes=32 time=63ms TTL=126

Reply from 192.168.40.1: bytes=32 time=78ms TTL=126

Reply from 192.168.40.1 bytes=32 time=94ms TTL=126

Reply from 192.168.40.1: bytes=32 time=78ms TTL=126

Ping statistics for 192.168.40.1:

    Packets: Sent = 4, Received = 4, Lost = 0 (0% loss),

Approximate round trip times in milli-seconds:

Minimum = 63ms, Maximum = 94ms, Average = 78ms

经过测试，PC2 无法和 pc3 通信，而和 PC1、PC4 的通信正常。俗话说"条条大路通罗马"，在网络实验中也是如此。解决问题的方式可能有多种，就看肯不肯动脑思考动手实践了。

项目 6

连入 Internet

我们平时所说的因特网，英文是 Internet，又被称为国际互联网，它是全球最大的信息资源库。当我们遇到各种疑难问题时，只要百度一下或在谷歌上搜一搜，瞬间就会找到答案。对于公司而言，在电子商务盛行的今天，接入因特网更是意义重大。公司不仅可以通过自己的网站宣传产品，进行网上交易等，还可以和远隔千山万水的分公司利用廉价的公用网络安全通信，好处真是不胜枚举。但一个公司的电脑数量往往在几十台到几百台，甚至上千台，而申请的公网 IP 数量很少。那么是如何保证数量众多的公司内网主机共享少量的公网 IP 接入 Internet 的呢？公司总部和分公司之间的通信又是如何跨越互联网像局域网一样安全方便呢？

通过本项目的实训，可以掌握静态 NAT、动态 NAT 和 NAPT 的配置，让我们共享网上冲浪的乐趣。同时，学会配置 IPSEC VPN 可以保证公司总部与分公司的安全通信。

任务 6.1 静态 NAT

【实训目的】

通过本实训，理解网络地址转换的原理和技术，掌握静态 NAT 的配置和测试。

【实训任务】

1. 根据拓扑图连接网络设备，构建实训网络。
2. 配置静态 NAT，实现使用私有 IP 访问因特网。
3. 验证 NAT 转化结果。

【预备知识】

一、内部地址和外部地址的相关概念

Inside local(内部本地地址)：一般情况下该地址是内部网络设备使用的私有地址。
inside global(内部全局地址)：一般是数据包从内网向外网转发时，由源 IP 转化成的公网 IP。
outside local(外部本地地址)： 内网设备或用户看到的外网设备的 IP 地址。
outside global(外部全局地址)：外部网络设备真实的 IP 地址。

二、静态 NAT 的配置

1. 指定内部接口

Router (config)#interface　接口类型　接口编号

Router（config-if）#ip　nat　inside

2. 指定外部接口

Router (config)#interface　接口类型　接口编号

Router（config-if）#ip　nat　outside

3. 配置并应用静态 NAT

Router（config）#ip　nat　inside　source　static　local-ip　global-ip

参数说明如下。

local-ip：内部本地地址。被转换的内网私有地址。

global-ip：内部全局地址。用来转换内部本地地址的公有 IP 地址。

【实训拓扑】

网络拓扑结构图如图 6-1 所示。

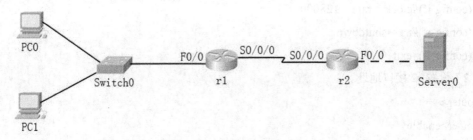

图 6-1 网络拓扑结构图

【实训设备】

路由器 2 台、二层交换机 1 台、计算机 3 台。

【实训步骤】

步骤 1 配置各设备的接口 IP 地址，参见表 6-1 所示。

表 6-1

设备名称	接口	IP	子网掩码
R1	F0/0	192.168.1.254	255.255.255.0
	S0/0/0	12.1.1.1	255.255.255.0
R2	S0/0/0	12.1.1.2	255.255.255.0
	F0/0	100.1.1.1	255.255.255.0
PC0	网卡	192.168.1.1	255.255.255.0
PC1	网卡	192.168.1.2	255.255.255.0
SERVER0	网卡	100.1.1.2	255.255.255.0

（1）设置 r1 接口地址。

Router>enable

Router#configure terminal

Router(config)#hostname r1

r1(config)#interface f0/0

r1(config-if)#ip address   192.168.1.254 255.255.255.0

r1(config-if)#no   shutdown

r1(config-if)#exit

r1(config)#interface s0/0/0

r1(config-if)#ip address 12.1.1.1 255.255.255.252

r1(config-if)#clock rate 128000

r1(config-if)#no shutdown

r1(config-if)#exit

（2）设置 r2 接口地址。

Router>

Router>enable

Router#configure terminal

Router(config)#hostname r2

r2(config)#interface s0/0/0

r2(config-if)#ip address 12.1.1.2 255.255.255.252

r2(config-if)#no shutdown

r2(config-if)#exit

r2(config)#

r2(config)#interface f0/0

r2(config-if)#ip address 100.1.1.1 255.255.255.0

r2(config-if)#no shutdown

r2(config-if)#exit

步骤 2 配置并检查各设备的路由。

（1）配置路由器 r1 的默认路由。

r1(config)#ip route 0.0.0.0 0.0.0.0 12.1.1.2

（2）查看路由器 r1 的路由信息。

r1#show ip route

Codes: C - connected, S - static, I - IGRP, R - RIP, M - mobile, B - BGP

      D - EIGRP, EX - EIGRP external, O - OSPF, IA - OSPF inter area

      N1 - OSPF NSSA external type 1, N2 - OSPF NSSA external type 2

      E1 - OSPF external type 1, E2 - OSPF external type 2, E - EGP

      i - IS-IS, L1 - IS-IS level-1, L2 - IS-IS level-2, ia - IS-IS inter area

      * - candidate default, U - per-user static route, o - ODR

      P - periodic downloaded static route

Gateway of last resort is 12.1.1.2 to network 0.0.0.0

    12.0.0.0/30 is subnetted, 1 subnets

C   12.1.1.0 is directly connected, Serial0/0/0

C   192.168.1.0/24 is directly connected, FastEthernet0/0

S*   0.0.0.0/0 [1/0] via 12.1.1.2

（3） 查看路由器 r2 的路由信息。

r2#show   ip   route

Codes: C - connected, S - static, I - IGRP, R - RIP, M - mobile, B - BGP

   D - EIGRP, EX - EIGRP external, O - OSPF, IA - OSPF inter area

   N1 - OSPF NSSA external type 1, N2 - OSPF NSSA external type 2

   E1 - OSPF external type 1, E2 - OSPF external type 2, E - EGP

   i - IS-IS, L1 - IS-IS level-1, L2 - IS-IS level-2, ia - IS-IS inter area

   * - candidate default, U - per-user static route, o - ODR

   P - periodic downloaded static route

Gateway of last resort is not set

  12.0.0.0/30 is subnetted, 1 subnets

C   12.1.1.0 is directly connected, Serial0/0/0

  100.0.0.0/24 is subnetted, 1 subnets

C   100.1.1.0 is directly connected, FastEthernet0/0

（4） 验证计算机 PC0、PC1 与 Server0 间的连通性。

①首先测试 PC0 和 Server0 的连通性：

PC>ping   100.1.1.2

Pinging 100.1.1.2 with 32 bytes of data:

Request timed out.

Request timed out.

Request timed out.

Request timed out.

Ping statistics for 100.1.1.2:

  Packets: Sent = 4, Received = 0, Lost = 4 (100% loss)

②再测试 PC1 和 Server0 的连通性：

PC>ping   100.1.1.2

Pinging 100.1.1.2 with 32 bytes of data:

Request timed out.

Request timed out.

Request timed out.

Request timed out.

Ping statistics for 100.1.1.2:

Packets: Sent = 4, Received = 0, Lost = 4 (100% loss)

以上输出结果显示，计算机 PC0 和 PC1 无法与 Server0 通信。原因是数据包从 PC0（PC1）发出后，经过路由器 R1 的路由转发到达路由器 R2，路由器 R2 将其送到直连的 Server0 上；应答的数据包从 Server0 到达路由器 R2 后，因为没有到达目的网络 192.168.1.0 的路由表项，故而将其丢弃。

步骤 3 在路由器 r1 上配置静态 NAT。

（1）设置内部转换接口。

r1(config)#interface　f0/0

r1(config-if)#ip　nat　inside

r1(config-if)#exit

（2）设置外部转换接口。

r1(config)#interface　s0/0/0

r1(config-if)#ip　nat　outside

r1(config-if)#exit

（3）应用 NAT。

r1(config)#ip　nat　inside　source　static　192.168.1.1　12.1.1.10

r1(config)#ip　nat　inside　source　static　192.168.1.2　12.1.1.20

（4）验证测试 PC0 和 Server0 的连通性。

PC>ping　100.1.1.2

Pinging 100.1.1.2 with 32 bytes of data:

Reply from 100.1.1.2: bytes=32 time=125ms TTL=126

Reply from 100.1.1.2: bytes=32 time=125ms TTL=126

Reply from 100.1.1.2: bytes=32 time=125ms TTL=126

Reply from 100.1.1.2: bytes=32 time=110ms TTL=126

Ping statistics for 100.1.1.2:

　　Packets: Sent = 4, Received = 4, Lost = 0 (0% loss),

Approximate round trip times in milli-seconds:

Minimum = 110ms, Maximum = 125ms, Average = 121ms

请同学们再测试 PC1 和 Server0 的互通性。

（5）查看路由器 r1 上的 NAT 转化结果。

r1#show ip　nat translations

Pro	Inside global	Inside local	Outside local	Outside global
---	12.1.1.10	192.168.1.1	---	---
---	12.1.1.20	192.168.1.2	---	---

为什么在我们做了静态 NAT 之后，PC0、PC1 就能够和 Server0 正常通信了呢？原来，数据包在路由器 R1 上路由转发前源 IP 地址被转变成了 12.1.1.10（12.1.1.20）。故而从 Server0 发回的应答数据包目的 IP 也是 12.1.1.10（12.1.1.20），路由器 R2 接收到包后从接口 s0/0/0 发出到直连的 12.1.1.0 网络上，路由器 R1 将收到数据包的目的包再转变成原来的 192.168.1.1（192.168.1.2）。

任务6.2 动态 NAT 和 NAPT

【实训目的】

通过本实训，理解网络地址转换的原理和技术，掌握动态 NAT 和 NAPT 的配置和测试。

【实训任务】

1. 根据拓扑图连接网络设备，构建实训网络。
2. 配置动态 NAT，实现使用私有 IP 访问因特网。
3. 验证 NAT 转化结果。

【预备知识】

一、动态 NAT 的配置

1. 指定内部接口

Router (config)#interface　接口类型　接口编号

Router（config-if）#ip　nat　inside

2. 指定外部接口

Router (config)#interface　接口类型　接口编号

Router（config-if）#ip　nat　outside

3. 定义 ACL，指明哪些范围内的主机可以进行 NAT

Router (config)#access-list　编号　permit　源网络号　源通配符掩码

如：Router (config)#access-list　10　permit　192.168.10.0 0.0.0.255

4. 定义内部全局地址池

Router (config)#ip　nat　pool　地址池名　内部全局起始地址　内部全局终止地址　netmask 子网掩码

假如我们定义地址池 internet，内有内部全局地址 100.1.1.1～100.1.1.5，子网掩码是 255.255.255.248。则相应的配置命令为

Router (config)#ip　nat　pool　internet　100.1.1.1　100.1.1.5　netmask　255.255.255.248

5. 配置应用动态 NAT

Router（config）#ip　nat　inside　source　list　编号　pool　地址池名

例如，我们调用前面第 3 步的 ACL 和第 4 步定义的地址池：

Router（config）#ip　nat　inside　source　list　10　pool　internet

二、NAPT 的配置

当访问公网的主机数多于拥有的公有 IP 数量时，我们就需要配置 NAPT。其配置过程和 NAT 相似，只是在最后应用 NAT 时添加选项 overload。即：

Router（config）#ip nat inside source list 编号 pool 地址池名 overload

示例：

Router（config）#ip nat inside source list 10 pool internet overload

【实训拓扑】

网络拓扑结构图如图 6-2 所示。

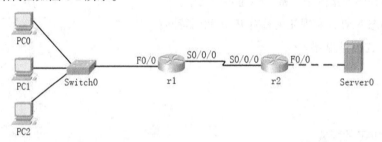

图 6-2 网络拓扑结构图

【实训设备】

路由器 2 台、二层交换机 1 台、计算机 4 台。

【实训步骤】

步骤 1 配置各设备的接口 IP 地址，参见表 6 -2 所示。

表 6-2

设备名称	接口	IP	子网掩码
r1	F0/0	192.168.1.254	255.255.255.0
	S0/0/0	12.1.1.1	255.255.255.0
r2	S0/0/0	12.1.1.2	255.255.255.0
	F0/0	100.1.1.1	255.255.255.0
PC0	网卡	192.168.1.1	255.255.255.0
PC1	网卡	192.168.1.2	255.255.255.0
PC2	网卡	192.168.1.3	255.255.255.0
Server0	网卡	100.1.1.2	255.255.255.0

（1）设置 R1 接口地址。

Router>enable

Router#configure terminal

Router(config)#hostname r1

r1(config)#interface f0/0

r1(config-if)#ip address 192.168.1.254 255.255.255.0

r1(config-if)#no shutdown

r1(config-if)#exit

r1(config)#interface s0/0/0

r1(config-if)#ip address 12.1.1.1 255.255.255.252

r1(config-if)#clock rate 128000

r1(config-if)#no shutdown

r1(config-if)#exit

（2）设置 R2 接口地址。

Router>

Router>enable

Router#configure terminal

Router(config)#hostname r2

r2(config)#interface s0/0/0

r2(config-if)#ip address 12.1.1.2 255.255.255.252

r2(config-if)#no shutdown

r2(config-if)#exit

r2(config)#

r2(config)#interface f0/0

r2(config-if)#ip address 100.1.1.1 255.255.255.0

r2(config-if)#no shutdown

r2(config-if)#exit

步骤 2　配置并检查各设备的路由。

（1）配置路由器 R1 的默认路由。

r1(config)#ip route 0.0.0.0 0.0.0.0 12.1.1.2

（2）查看路由器 R1 的路由信息。

r1#show ip route

Codes: C - connected, S - static, I - IGRP, R - RIP, M - mobile, B - BGP

　　　 D - EIGRP, EX - EIGRP external, O - OSPF, IA - OSPF inter area

　　　 N1 - OSPF NSSA external type 1, N2 - OSPF NSSA external type 2

　　　 E1 - OSPF external type 1, E2 - OSPF external type 2, E - EGP

i - IS-IS, L1 - IS-IS level-1, L2 - IS-IS level-2, ia - IS-IS inter area

   * - candidate default, U - per-user static route, o - ODR

   P - periodic downloaded static route

Gateway of last resort is 12.1.1.2 to network0.0.0.0

   12.0.0.0/30 is subnetted, 1 subnets

C 12.1.1.0 is directly connected, Serial0/0/0

C 192.168.1.0/24 is directly connected, FastEthernet0/0

S* 0.0.0.0/0 [1/0] via 12.1.1.2

（3）查看路由器 R2 的路由信息。

r2#show ip route

Codes: C - connected, S - static, I - IGRP, R - RIP, M - mobile, B - BGP

   D - EIGRP, EX - EIGRP external, O - OSPF, IA - OSPF inter area

   N1 - OSPF NSSA external type 1, N2 - OSPF NSSA external type 2

   E1 - OSPF external type 1, E2 - OSPF external type 2, E - EGP

   i - IS-IS, L1 - IS-IS level-1, L2 - IS-IS level-2, ia - IS-IS inter area

   * - candidate default, U - per-user static route, o - ODR

   P - periodic downloaded static route

Gateway of last resort is not set

   12.0.0.0/30 is subnetted, 1 subnets

C 12.1.1.0 is directly connected, Serial0/0/0

   100.0.0.0/24 is subnetted, 1 subnets

C 100.1.1.0 is directly connected, FastEthernet0/0

（4）验证计算机 PC0、PC1 与 SERVER0 间的连通性。

首先测试 PC0 和 Server0 的连通性：

PC>ping 100.1.1.2

Pinging 100.1.1.2 with 32 bytes of data:

Request timed out.

Request timed out.

Request timed out.

Request timed out.

Ping statistics for 100.1.1.2:

  Packets: Sent = 4, Received = 0, Lost = 4 (100% loss)

再测试 PC1 和 SERVER0 的连通性：

PC>ping 100.1.1.2

Pinging 100.1.1.2 with 32 bytes of data:

Request timed out.

Request timed out.

Request timed out.

Request timed out.

Ping statistics for 100.1.1.2:

Packets: Sent = 4, Received = 0, Lost = 4 (100% loss)

以上输出结果显示，计算机 PC0 和 PC1 无法与 Server0 通信。原因是数据包从 PC0（PC1）发出后，经过路由器 R1 的路由转发到达路由器 R2，路由器 R2 将其送到直连的 Server0 上；应答的数据包从 Server0 到达路由器 R2 后，因为没有到达目的网络 192.168.1.0 的路由表项，故而将其丢弃。

步骤 3 在路由器 R1 上配置动态 NAT。

（1）设置内部转换接口。

r1(config)#interface   f0/0

r1(config-if)#ip   nat   inside

r1(config-if)#exit

（2）设置外部转换接口。

r1(config)#interface   s0/0/0

r1(config-if)#ip   nat   outside

r1(config-if)#exit

（3）定义 ACL。

r1(config)#access-list   1   permit   192.168.1.0 0.0.0.255

（4）定义内部全局地址池。

r1(config)#ip   nat   pool   internet   192.168.1.100   192.168.1.101   netmask   255.255.255.0

（5）应用动态 NAT。

r1(config)#ip   nat   inside source list   1   pool   internet

步骤 4 验证测试计算机 pc0 和 server0 的连通性。

PC>ping 100.1.1.2

Pinging 100.1.1.2 with 32 bytes of data:

Reply from 100.1.1.2: bytes=32 time=125ms TTL=126

Reply from 100.1.1.2: bytes=32 time=125ms TTL=126

Reply from 100.1.1.2: bytes=32 time=125ms TTL=126

Reply from 100.1.1.2: bytes=32 time=125ms TTL=126

Ping statistics for 100.1.1.2:

　　Packets: Sent = 4, Received = 4, Lost = 0 (0% loss),

Approximate round trip times in milli-seconds:

Minimum = 125ms, Maximum = 125ms, Average = 125ms

步骤 5 观察路由器 R1 上的 NAT 结果。

r1#show ip nat translations
Pro   Inside global   Inside local   Outside local   Outside global
icmp   12.1.1.100:5   192.168.1.1:5   100.1.1.2:5   100.1.1.2:5

以上输出结果显示，pc0 和 server0 通信时，IP 地址由内部局部地址 192.168.1.1 转变为内部全局地址 12.1.1.100。

步骤 6 测试 pc1、pc2 与 server0 间的连通性。

（1）用 pc1 去 ping server0：

PC>ping 100.1.1.2

Pinging 100.1.1.2 with 32 bytes of data:

Reply from 100.1.1.2: bytes=32 time=125ms TTL=126

Reply from 100.1.1.2: bytes=32 time=125ms TTL=126

Reply from 100.1.1.2: bytes=32 time=125ms TTL=126

Reply from 100.1.1.2: bytes=32 time=125ms TTL=126

Ping statistics for 100.1.1.2:

　　Packets: Sent = 4, Received = 4, Lost = 0 (0% loss),

Approximate round trip times in milli-seconds:

Minimum = 125ms, Maximum = 125ms, Average = 125ms

（2）用 pc2 去 ping server0：

PC>ping 100.1.1.2

Pinging 100.1.1.2 with 32 bytes of data:

Request timed out.

Request timed out.

Request timed out.

Request timed out.

Ping statistics for 100.1.1.2:

　　Packets: Sent = 4, Received = 0, Lost = 4 (100% loss),

经过观察可以发现：pc1 能 ping 通 server0，而 pc2 却不可以。为什么呢？我们观察路由器 R1 上的 NAT 结果：

r1#show ip nat translations
Pro   Inside global   Inside local   Outside local   Outside global
icmp   12.1.1.100:6   192.168.1.1:6   100.1.1.2:6   100.1.1.2:6
icmp   12.1.1.101:2   192.168.1.2:2   100.1.1.2:2   100.1.1.2:2

原来，众多的内网主机争用仅有的两个公有 IP，只有最早上线的两台主机才能获得访问 internetwork 的资格。那么其他主机就没有资格共享上网访问因特网吗？我们可以用 NAPT 技术解决上述问题。

步骤 7　应用 NAPT。

（1）撤销 NAT。

r1(config)#no ip nat inside source list 1

（2）应用 NAPT。

r1(config)#ip nat inside source list 1 pool internet overload

（3）测试 pc0、pc1、pc2 与 server0 间的连通性，观察 NAT 结果。

首先检验 pc0 和 server0 的连通性：

PC>ping 100.1.1.2

Pinging 100.1.1.2 with 32 bytes of data:

Reply from 100.1.1.2: bytes=32 time=80ms TTL=127

Reply from 100.1.1.2: bytes=32 time=94ms TTL=127

Reply from 100.1.1.2: bytes=32 time=94ms TTL=127

Reply from 100.1.1.2: bytes=32 time=94ms TTL=127

Ping statistics for 100.1.1.2:

  Packets: Sent = 4, Received = 4, Lost = 0 (0% loss),

Approximate round trip times in milli-seconds:

  Minimum = 80ms, Maximum = 94ms, Average = 90ms

然后，再检验 pc1 和 server0 的连通性：

PC>ping 100.1.1.2

Pinging 100.1.1.2 with 32 bytes of data:

Reply from 100.1.1.2: bytes=32 time=93ms TTL=127

Reply from 100.1.1.2: bytes=32 time=94ms TTL=127

Reply from 100.1.1.2: bytes=32 time=94ms TTL=127

Reply from 100.1.1.2: bytes=32 time=93ms TTL=127

Ping statistics for 100.1.1.2:

  Packets: Sent = 4, Received = 4, Lost = 0 (0% loss),

Approximate round trip times in milli-seconds:

  Minimum = 93ms, Maximum = 94ms, Average = 93ms

接着，检验 pc2 和 server0 的连通性：

PC>ping 100.1.1.2

Pinging 100.1.1.2 with 32 bytes of data:

Reply from 100.1.1.2: bytes=32 time=125ms TTL=126

Reply from 100.1.1.2: bytes=32 time=125ms TTL=126

Reply from 100.1.1.2: bytes=32 time=125ms TTL=126

Reply from 100.1.1.2: bytes=32 time=125ms TTL=126

Ping statistics for 100.1.1.2:

  Packets: Sent = 4, Received = 4, Lost = 0 (0% loss),

Approximate round trip times in milli-seconds:

Minimum = 125ms, Maximum = 125ms, Average = 125ms

结果证明：计算机 pc0、pc1、pc2 均能访问 server0。

最后，观察 NAT 转化结果：

r1#show ip nat translations

Pro	Inside global	Inside local	Outside local	Outside global
icmp	12.1.1.101:9	192.168.1.1:9	100.1.1.2:9	100.1.1.2:9
icmp	12.1.1.101:5	192.168.1.2:5	100.1.1.2:5	100.1.1.2:5
icmp	12.1.1.101:4	192.168.1.3:4	100.1.1.2:4	100.1.1.2:4

由此不难得出结论：多台内部主机共享少量的公有 IP 接入公网，可以用 NAPT 技术分配不同的端口号加以区分各主机。由此可见，overload 十分重要，请同学们今后实训时切记应用这个选项。

任务6.3 IPSEC VPN

【实训目的】

掌握 IPSEC VPN 的配置，构建安全的互联网环境。

【实训任务】

1. 配置 IPSEC VPN。
2. 验证测试 IPSEC VPN。

【预备知识】

一、启用 IKE 协商

Roue(config)#crypto isakmp policy 10 //建立 IKE 协商策略（10 是策略编号，数值越小优先级越高。）

Route(config-isakmp)#hash md5 //认证密钥的算法为 MD5

Route(config-isakmp)#authentication pre-share//告知路由器使用预先共享的密钥

Route(config-isakmp)#exit

Route(config)#crypto isakmp key CISCO address 10.0.0.6// CISCO 为预共享密钥，对端预共享密钥也必须是 CISCO，10.0.0.6 为对端路由器地址

二、配置 IPSEC 参数

Route(config)#crypto ipsec transform-set ipsecvpn ah-md5-hmac esp-des//ipsecvpn 为 transform-set 的名称。ah-md5-hamc esp-des 是采用的验证和加密算法

Route(config)#access-list 101 permit ip 192.168.1.0 0.0.0.255 200.1.1.0 0.0.0.255//扩展访问控制列表。192.168.1.0 段 ~ 200.1.1.0 段的数据需要封装保护

三、设置 crypto map

Route(config)#crypto map vpn 10 ipsec-isakmp//vpn: crypto map 的名称。10：优先级。ipsec-isakmp 表示此 IPSec 链接采用 IKE 自动协商

Route(config-crypto-map)#set peer 10.0.0.6//指定此 VPN 链路对端地址为 10.0.0.6

Route(config-crypto-map)#set transform-set ipsecvpn//调用配置的 transform-set

Route(config-crypto-map)#match address 101//匹配访问控制列表 101。

四、把 crypto map 应用到端口

Route(config)#interface fastEthernet 0/0 //进入 F0/0 端口

Route(config-if)#crypto map vpn //把 crypto map 应用到端口

【实训拓扑】

网络拓扑结构图如图 6-3 所示。

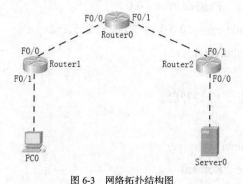

图 6-3 网络拓扑结构图

【实训设备】

路由器 3 台、计算机 2 台。

【实训步骤】

步骤 1 根据拓扑图构建好网络，配置计算机的 IP 地址，参见表 6-3-1 所示。

表 6-3-1

设备名称	接口	IP 地址	子网掩码	网关
PC0	网卡	192.168.1.2	255.255.255.0	192.168.1.1
Server0	网卡	200.1.1.2	255.255.255.0	200.1.1.1
Router0	F0/0	10.0.0.2	255.255.255.252	
	F0/1	10.0.0.5	255.255.255.252	
Router1	F0/0	10.0.0.1	255.255.255.252	
	F0/1	192.168.1.1	255.255.255.0	
Router2	F0/0	200.1.1.1	255.255.255.0	
	F0/1	10.0.0.6	255.255.255.252	

步骤 2 配置 Router1、Router0、Router2 的接口 IP 并开启。

（1）配置 Router1 的接口 IP 并开启。

Router>enable

Router#configure terminal

Router (config)#hostname Router1

Router1(config)#interface FastEthernet 0/1

Router1(config-if)#ip address 192.168.1.1 255.255.255.0

Router1(config-if)#no shutdown

Router1(config-if)#exit

Router1(config)#interface FastEthernet 0/0

Router1(config-if)#ip address 10.0.0.1 255.255.255.252

Router1(config-if)#no shutdown

Router1(config-if)#exit

（2）配置 Router0 的接口 IP 并开启。

Router>enable

Router#configure terminal

Router(config)#hostname Router0

Router0(config)#interface FastEthernet 0/0

　Router0(config-if)#ip address 10.0.0.2 255.255.255.252

Router0(config-if)#no shutdown

Router0(config-if)#exit

Router0(config)#interface FastEthernet 0/1

　Router0(config)#ip address 10.0.0.5 255.255.255.252

Router0(config-if)#no shutdown

Router0(config-if)#exit

（3）配置 Router2 的接口 IP 并开启。

Router>enable

Router#configure terminal

Router(config)#hostname Router2

Router2(config)#interface FastEthernet 0/0

Router2(config-if)#ip address 200.1.1.1 255.255.255.0

Router2(config-if)#no shutdown

Router2(config-if)#exit

Router2(config)#interface FastEthernet 0/1

Router2(config-if)#ip address 10.0.0.6 255.255.255.252

Router2(config-if)#no shutdown

Router2(config-if)#exit

步骤 3 在 Router1 和 Router2 上配置默认路由。

（1）在 Router1 上配置默认路由。

Router1(config)#ip route 0.0.0.0 0.0.0.0 10.0.0.2

（2）在 Router2 上配置默认路由。

Router2(config)#ip route 0.0.0.0 0.0.0.0 10.0.0.5

步骤 4 在 Router1 上配置并应用 IPSEC VPN。

（1）在 Router1 上启用 IKE 协商。

Router1(config)#crypto isakmp policy 10

Route1(config-isakmp)#hash md5

Route1(config-isakmp)#authentication pre-share

Route1(config-isakmp)#exit

Route1(config)#crypto isakmp key cisco address 10.0.0.6

（2）在 Router1 上配置 IPSEC 参数。

Route1(config)#crypto ipsec transform-set vpn esp-3des esp-md5-hmac

Route1(config)#access-list 110 permit ip 192.168.1.0 0.0.0.255 200.1.1.0 0.0.0.255

（3）在 Router1 上设置 crypto map。

Route1(config)#crypto map mymap 10 ipsec-isakmp

Route1(config-crypto-map)#set peer 10.0.0.6

Route1(config-crypto-map)#set transform-set vpn

Route1(config-crypto-map)#match address 110

Route1(config-crypto-map)#exit

（4）把 crypto map 应用到端口 f0/0。

Route1(config)#interface FastEthernet 0/0

Router1(config-if)#crypto map mymap

Router1(config-if)#exit

步骤 5 在 Router2 上配置并应用 IPSEC VPN。

（1）在 Router2 上启用 IKE 协商。

Router2(config)#crypto isakmp policy 10

  Route2(config-isakmp)#hash md5

  Route2(config-isakmp)#authentication pre-share

Route2(config-isakmp)#exit

Route2(config)#crypto isakmp key cisco address 10.0.0.1

（2）在 Router2 上配置 IPSEC 参数。

Route2(config)#crypto ipsec transform-set vpn esp-3des esp-md5-hmac

Route2(config)# access-list 110 permit ip 200.1.1.0 0.0.0.255 192.168.1.0 0.0.0.255

（3）在 Router2 上设置 crypto map。

Route2(config)#crypto map mymap 10 ipsec-isakmp

  Route2(config-crypto-map)#set peer 10.0.0.1

  Route2(config-crypto-map)#set transform-set vpn

  Route2(config-crypto-map)#match address 110

Route2(config-crypto-map)#exit

（4）应用 crypto map 到端口 f0/1。

Route2(config)#interface FastEthernet 0/0

Router2(config-if)#crypto map mymap

Router2(config-if)#exit

步骤 6 验证测试。

使用 Pc0 去 ping 服务器 Server0：

PC>ping -t 200.1.1.2

Pinging 200.1.1.2 with 32 bytes of data:

Request timed out.

Reply from 200.1.1.2: bytes=32 time=125ms TTL=126

Reply from 200.1.1.2: bytes=32 time=109ms TTL=126

Reply from 200.1.1.2: bytes=32 time=125ms TTL=126

Reply from 200.1.1.2: bytes=32 time=125ms TTL=126

Reply from 200.1.1.2: bytes=32 time=125ms TTL=126

Reply from 200.1.1.2: bytes=32 time=125ms TTL=126

Reply from 200.1.1.2: bytes=32 time=125ms TTL=126

Reply from 200.1.1.2: bytes=32 time=125ms TTL=126

Reply from 200.1.1.2: bytes=32 time=125ms TTL=126

Reply from 200.1.1.2: bytes=32 time=125ms TTL=126

Ping statistics for 200.1.1.2:

    Packets: Sent = 11, Received = 10, Lost = 1 (10% loss),

Approximate round trip times in milli-seconds:

    Minimum = 109ms, Maximum = 125ms, Average = 123ms

经过观察，在刚开始 ping 时会出现丢包，原因是 IKE 协商尚未成功。待协商成功后，数据被封装经过 IPSEC 隧道到达目的站点。

Router1#show crypto isakmp sa

IPv4 Crypto ISAKMP SA

dst	src	state	conn-id	slot	status
10.0.0.6	10.0.0.1	QM_IDLE	1071	0	ACTIVE

表明从源 10.0.0.1 到目的 10.0.0.6 的 IKE 协商成功。

Router#show crypto ipsec sa

interface: FastEthernet0/0

    Crypto map tag: mymap, local addr 10.0.0.1

  protected vrf: (none)

  local   ident (addr/mask/prot/port): (192.168.1.0/255.255.255.0/0/0)

  remote   ident (addr/mask/prot/port): (200.1.1.0/255.255.255.0/0/0)

  current_peer 10.0.0.6 port 500

    PERMIT, flags={origin_is_acl,}

  #pkts encaps: 10, #pkts encrypt: 10, #pkts digest: 0

  #pkts decaps: 10, #pkts decrypt: 10, #pkts verify: 0

  #pkts compressed: 0, #pkts decompressed: 0

  #pkts not compressed: 0, #pkts compr- failed: 0

  #pkts not decompressed: 0, #pkts decompress failed: 0

  #send errors 1, #recv errors 0

    local crypto endpt-: 10.0.0.1, remote crypto endpt-:10.0.0.6

    path mtu 1500, ip mtu 1500, ip mtu idb FastEthernet0/0

    current outbound spi: 0x0AB21023(179441699)

    inbound esp sas:

     spi: 0x0A334ED5(171134677)

       transform: esp-3des esp-md5-hmac ,

       in use settings ={Tunnel, }

       conn id: 2009, flow_id: FPGA:1, crypto map: mymap

       sa timing: remaining key lifetime (k/sec): (4525504/3415)

       IV size: 16 bytes

       replay detection support: N

Status: ACTIVE　　//表示输入流量的安全关联建立成功
   inbound ah sas:
   inbound pcp sas:
  outbound esp sas:
   spi: 0x0AB21023(179441699)
    transform: esp-3des esp-md5-hmac ,
    in use settings ={Tunnel, }
    conn id: 2010, flow_id: FPGA:1, crypto map: mymap
    sa timing: remaining key lifetime (k/sec): (4525504/3415)
    IV size: 16 bytes
    replay detection support: N
    Status: ACTIVE　　//表示输出流量的安全关联建立成功
 outbound ah sas:
  outbound pcp sas:
IPSEC VPN 常用的查看命令如表 6-3-2 所示。

表 6-3-2

show crypto map	查看加密映射图
show crypto isakmp policy	查看密钥交换策略
Show crypto isakmp key	查看当前密钥交换方式所使用的密钥
show crypto isakmp peers	查看已建立的对等体
show crypto isakmp sa	查看安全关联
show crypto ipsec transform-set	查看 IPSec 加密转换集

任务 6.4　配置 IPSEC VPN 和 NAT

【实训目的】

  配置 IPSEC VPN 和 NAT，实现 IPSEC VPN 站点到站点的远程连接，同时利用 NAT 实现连接 Internet。

【实训任务】

  1. 配置 IPSEC VPN。
  2. 配置 NAT。

【预备知识】

当我们在同一出口路由器上同时配置 IPSEC VPN 和 NAT 时，由于首先进行 NAT 转换，要通过 VPN 保护的数据包源 IP 被转化为公网 IP，从而使 IPSEC VPN 失效。所以，定义 NAT 转换所需的 ACL 就成为关键所在。

NAT 转换需要的扩展访问列表（编号为 101～199）：

Router(config)#access-list 编号 deny ip 源私有网段 反掩码 目的私有网段 反掩码
Router(config)#access-list 编号 permit ip any any

例如，总公司内网网段 192.168.10.0/24，分公司内网网段 172.16.10.0/24，则我们在总公司出口路由器上定义 NAT 转换用到的扩展 ACL 配置如下：

Router(config)#access-list 110 deny ip 192.168.10.0 0.0.0.255 172.16.10.0 0.0.0.255
Router(config)#access-list 110 permit ip any any

在分公司出口路由器上定义 NAT 转换用到的扩展 ACL 配置如下：

Router(config)#access-list 110 deny ip 172.16.10.0 0.0.0.255 192.168.10.0 0.0.0.255
Router(config)#access-list 110 permit ip any any

【实训拓扑】

网络拓扑结构图如图 6-4 所示。

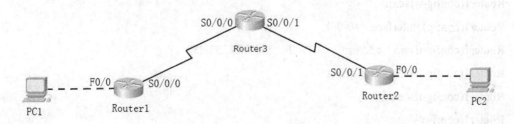

图 6-4 网络拓扑结构图

【实训设备】

路由器 3 台、计算机 2 台。

【实训步骤】

步骤 1 按照拓扑图搭建网络，配置各路由器接口 IP 地址和计算机 IP 地址，参见表 6-4 所示。

表 6-4

设备名称	接口	IP 地址	子网掩码	网关
PC1	网卡	192.168.1.2	255.255.255.0	192.168.1.1
PC2	网卡	172.16.1.2	255.255.0.0	172.16.1.1
Router1	F0/0	192.168.1.1	255.255.255.0	
	S0/0/0	13.1.1.1	255.255.255.252	
Router2	F0/0	172.16.1.1	255.255.0.0	
	S0/0/1	23.1.1.1	255.255.255.252	
Router3	S0/0/0	13.1.1.2	255.255.255.252	
	S0/0/1	23.1.1.2	255.255.255.252	

（1）配置路由器 Router1。

Router>enable

Router#configure terminal

Router(config)#hostname Router1

Router1(config)#interface f0/0

Router1(config-if)#ip address 192.168.1.1 255.255.255.0

Router1(config-if)#no shutdown

Router1(config-if)#exit

Router1(config)#interface s0/0/0

Router1(config-if)#ip address 13.1.1.1 255.255.255.252

Router1(config-if)#no shutdown

Router1(config-if)#exit

Router1(config)#

（2）配置路由器 Router3。

Router>enable

Router#configure terminal

Router(config)#hostname Router3

Router3(config)#interface s0/0/0

Router3(config-if)#ip address 13.1.1.2   255.255.255.252

Router3(config-if)#clock rate 128000

Router3(config-if)#no shutdown

Router3(config-if)#exit

Router3(config)#interface s0/0/1

Router3(config-if)#ip address 23.1.1.2 255.255.255.252

Router3(config-if)#clock rate 128000
Router3(config-if)#no shutdown
Router3(config-if)#end
Router3#

(3)配置路由器 Router2。

Router>enable
Router#configure terminal
Router(config)#hostname Router2
Router2(config)#interface f0/0
Router2(config-if)#ip address 172.16.1.1 255.255.0.0
Router2(config-if)#no shutdown
Router2(config-if)#exit
Router2(config)#interface s0/0/1
Router2(config-if)#ip address 23.1.1.1 255.255.255.252
Router2(config-if)#no shutdown
Router2(config-if)#exit
Router2(config)#

步骤 2 配置路由器 Router1 和 Router2 的默认路由。

(1)配置路由器 Router1。

Router1(config)#ip route 0.0.0.0 0.0.0.0 13.1.1.2

(2)配置路由器 Router2。

Router2(config)#ip route 0.0.0.0 0.0.0.0 23.1.1.2

步骤 3 配置 IPSEC VPN。

(1)配置路由器 Router1。

Router1(config)#crypto isakmp policy 10 //建立 IKE 协商策略
Router1(config-isakmp)#authentication pre-share
Router1(config-isakmp)#hash md5
Router1(config-isakmp)#exit
Router1(config)#crypto isakmp key cisco address 23.1.1.1//设置共享密钥为 cisco,对端地址为 23.1.1.1
Router1(config)#crypto ipsec transform-set vpn esp-3des esp-md5-hmac
　　　　　　　　//配置转换集 vpn 及其加密算法。
Router1(config)#access-list 100 permit ip 192.168.1.0 0.0.0.255 172.16.0.0 0.0.255.255
　　　　　　　　//定义 IPSEC VPN 要保护的数据流。
Router1(config)#crypto map ipsec-vpn 10 ipsec-isakmp
Router1(config-crypto-map)#set peer 23.1.1.1 //设置 VPN 链路对端地址为 23.1.1.1。

Router1(config-crypto-map)#set transform-set vpn   //设置采用的转换集名为 vpn

Router1(config-crypto-map)#match address 100   //匹配要保护的数据流符合 acl 100 规则

Router1(config-crypto-map)#exit

Router1(config)#interface s0/0/0

Router1(config-if)#crypto map ipsec-vpn   //在接口上应用 ipsec-vpn。

Router1(config-if)#exit

Router1(config)#

（2）配置路由器 Router2。

Router2(config)#crypto isakmp policy 10

Router2(config-isakmp)#authentication pre-share

Router2(config-isakmp)#hash md5

Router2(config-isakmp)#exit

Router2(config)#crypto isakmp key cisco address 12.1.1.1

Router2(config)#crypto ipsec transform-set vpn esp-3des esp-md5-hmac

Router2(config)#access-list 100 permit ip 172.16.0.0 0.0.255.255 192.168.1.0 0.0.0.255

Router2(config)#crypto map ipsec-vpn 10 ipsec-isakmp

Router2(config-crypto-map)#set peer 12.1.1.1

Router2(config-crypto-map)#set transform-set vpn

Router2(config-crypto-map)#match address 100

Router2(config-crypto-map)#exit

Router2(config)#interface s0/0/1

Router2(config-if)#crypto map ipsec-vpn

Router2(config-if)#exit

Router2(config)#

（3）验证 IPSEC VPN。

在 PC1 上运行 ping 命令：

PC>ping 172.16.1.2

Pinging 172.16.1.2 with 32 bytes of data:

Request timed out.

Reply from 172.16.1.2: bytes=32 time=110ms TTL=126

Reply from 172.16.1.2: bytes=32 time=125ms TTL=126

Reply from 172.16.1.2: bytes=32 time=112ms TTL=126

Ping statistics for 172.16.1.2:

　　Packets: Sent = 4, Received = 3, Lost = 1 (25% loss),

Approximate round trip times in milli-seconds:

Minimum = 110ms, Maximum = 125ms, Average = 115ms

结果为 ping 通,表明 VPN 配置成功。

接下来,我们使用 show 命令查看 crypto 状态:

Router1#show crypto isakmp sa

IPv4 Crypto ISAKMP SA

dst	src	state	conn-id slot status
23.1.1.1	13.1.1.1	QM_IDLE	1076 0 ACTIVE //IKE 协商成功

IPv6 Crypto ISAKMP SA

Router1#show crypto ipsec sa

interface: Serial0/0/0

    Crypto map tag: ipsec-vpn, local addr 13.1.1.1

  protected vrf: (none)

  local　ident (addr/mask/prot/port): (192.168.1.0/255.255.255.0/0/0)

  remote　ident (addr/mask/prot/port): (172.16.0.0/255.255.0.0/0/0)

  current_peer 23.1.1.1 port 500

    PERMIT, flags={origin_is_acl,}

  #pkts encaps: 253, #pkts encrypt: 253, #pkts digest: 0

  #pkts decaps: 253, #pkts decrypt: 253, #pkts verify: 0

  #pkts compressed: 0, #pkts decompressed: 0

  #pkts not compressed: 0, #pkts compr- failed: 0

  #pkts not decompressed: 0, #pkts decompress failed: 0

  #send errors 1, #recv errors 0

    local crypto endpt-: 13.1.1.1, remote crypto endpt-:23.1.1.1

    path mtu 1500, ip mtu 1500, ip mtu idb Serial0/0/0

    current outbound spi: 0x60BD6256(1623024214)

    inbound esp sas:

     spi: 0x477979F7(1199143415)

       transform: esp-3des esp-md5-hmac ,

       in use settings ={Tunnel, }

       conn id: 2004, flow_id: FPGA:1, crypto map: ipsec-vpn

       sa timing: remaining key lifetime (k/sec): (4525504/3312)

       IV size: 16 bytes

       replay detection support: N

       Status: ACTIVE　　//表示输入流量的安全关联建立成功

    inbound ah sas:

    inbound pcp sas:

    outbound esp sas:

spi: 0x60BD6256(1623024214)

 transform: esp-3des esp-md5-hmac ,

 in use settings ={Tunnel, }

 conn id: 2005, flow_id: FPGA:1, crypto map: ipsec-vpn

 sa timing: remaining key lifetime (k/sec): (4525504/3312)

 IV size: 16 bytes

 replay detection support: N

 Status: ACTIVE   //表示输出流量的安全关联建立成功

outbound ah sas:

outbound pcp sas:

Router1#

步骤 4 配置 NAT。

（1）配置路由器 Router1。

Router1#conf terminal

Router1(config)#access-list   110   deny   ip 192.168.1.0 0.0.0.255   172.16.0.0   0.0.255.255

//本条命令最关键，禁止前往 172.16.0.0 网段的源 IP 进行 NAT

Router1(config)#access-list   110   permit   ip   any   any

Router1(config)#interface   f0/0

Router1(config-if)#ip   nat   inside

Router1(config-if)#exit

Router1(config)#interface   s0/0/0

Router1(config-if)#ip   nat   outside

Router1(config-if)#exit

Router1(config)#ip   nat   inside   source   list   110   interface   s0/0/0   overload

Router1(config)#

（2）配置路由器 Router2。

Router2#configure   terminal

Router2(config)#access-list   110   deny   ip   172.16.0.0   0.0.255.255   192.168.1.0 0.0.0.255

//本条命令最关键，禁止前往 192.168.1.0 网段的源 IP 进行 NAT

Router2(config)#access-list   110   permit   ip   any   any

Router2(config)#interface   f0/0

Router2(config-if)#ip   nat   inside

Router2(config-if)#exit

Router2(config)#interface   s0/0/1

Router2(config-if)#ip   nat   outside

Router2(config-if)#exit

Router2(config)#ip nat inside source list 110 interface s0/0/1 overload
Router2(config)#

步骤 5 验证测试。

在 PC1 上 ping 对方私有网段计算机 PC2：

PC>ping   172.16.1.2

Pinging 172.16.1.2 with 32 bytes of data:

Reply from 172.16.1.2: bytes=32 time=125ms TTL=126

Reply from 172.16.1.2: bytes=32 time=125ms TTL=126

Reply from 172.16.1.2: bytes=32 time=109ms TTL=126

Reply from 172.16.1.2: bytes=32 time=125ms TTL=126

Ping statistics for 172.16.1.2:

　　Packets: Sent = 4, Received = 4, Lost = 0 (0% loss),

Approximate round trip times in milli-seconds:

　　Minimum = 109ms, Maximum = 125ms, Average = 121ms

结果为 ping 通。然后在路由器 Router1 上查看 NAT 情况：

Router1#show   ip   nat   translations

从 PC1 发出到 PC2 的数据包的源 IP（192.168.1.2）没有进行 NAT 转化，而是通过 VPN 进行通信。再用 PC1 去 ping 公网中路由器 Router3 的接口 IP：13.1.1.2。

PC>ping   13.1.1.2

Pinging 13.1.1.2 with 32 bytes of data:

Reply from 13.1.1.2: bytes=32 time=63ms TTL=254

Reply from 13.1.1.2: bytes=32 time=63ms TTL=254

Reply from 13.1.1.2: bytes=32 time=63ms TTL=254

Reply from 13.1.1.2: bytes=32 time=62ms TTL=254

Ping statistics for 13.1.1.2:

　　Packets: Sent = 4, Received = 4, Lost = 0 (0% loss),

Approximate round trip times in milli-seconds:

　　Minimum = 62ms, Maximum = 63ms, Average = 62ms

仍然正常通信。我们在路由器 Router1 上查看 NAT 情况：

Router1#show   ip   nat   translations

Pro	Inside global	Inside local	Outside local	Outside global
icmp	13.1.1.1:351	192.168.1.2:351	13.1.1.2:351	13.1.1.2:351
icmp	13.1.1.1:352	192.168.1.2:352	13.1.1.2:352	13.1.1.2:352
icmp	13.1.1.1:353	192.168.1.2:353	13.1.1.2:353	13.1.1.2:353
icmp	13.1.1.1:354	192.168.1.2:354	13.1.1.2:354	13.1.1.2:354

结果表明：从 PC1 到因特网的数据包源 IP 转化成公网 IP13.1.1.1。

这样，我们既保证了两个私有网段可以通过 IPSEC VPN 进行像在专用的私有网络中一样的安全通信，同时，又保证了私有网段内的主机正常连接互联网。

任务6.5 配置 GRE VPN

【实训目的】

配置 GRE VPN，实现站点到站点的远程连接。

【实训任务】

1. 配置 GRE VPN。
2. 配置静态路由。
3. 验证测试 GRE VPN。

【预备知识】

一、什么是 GRE

GRE（Generic Routing Encapsulation）即通用路由封装协议，是对某些网络层协议(如 IP 和 IPX)的数据包进行封装，使这些被封装的数据包能够在另一个网络层协议(如 IP)中传输。

GRE 是 VPN(Virtual Private Network)的第三层隧道协议，即在协议层之间采用了一种被称之为 Tunnel(隧道)的技术。

二、GRE 的特点

GRE 是一个标准协议，支持多种协议和多播，能够用来创建弹性的 VPN，并支持多点隧道。但是，GRE 缺乏加密机制，往往结合各种安全措施来保护数据，如利用 Ipsec 对其数据加密。

三、GRE VPN 原理

GRE 是 VPN 技术的一种。它在私网 IP 包头前再封装一个公网的 IP 包头，路由器根据公网 IP 进行路由转发，当目标路由器接收到数据后，去除公网 IP 包头，发现是 GRE 数据，然后将私网 IP 数据传送到内网中。

【实训拓扑】

网络拓扑结构图如图 6-5-1 所示。

项目6 连入 Internet

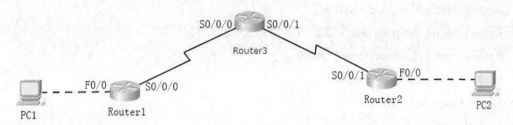

图 6-5-1 网络拓扑结构图

【实训设备】

路由器 3 台、计算机 2 台。

【实训步骤】

步骤 1 按照拓扑图搭建网络，并配置各设备 IP 地址，参见表 6-5 所示。

表 6-5

设备名称	接口	IP	子网掩码	网关
Router1	S0/0/0	12.1.1.1	255.255.255.252	
	F0/0	192.168.10.1	255.255.255.0	
Router2	S0/0/1	23.1.1.1	255.255.255.252	
	F0/0	192.168.20.1	255.255.255.0	
Router3	S0/0/0	12.1.1.2	255.255.255.252	
	S0/0/0	23.1.1.2	255.255.255.252	
PC1	网卡	192.168.10.10	255.255.255.0	192.168.10.1
PC2	网卡	192.168.20.20	255.255.255.0	192.168.20.1

（1）配置 Router1。

Router>enable

Router#configure Terminal

Enter configuration commands, one per line.　End with CNTL/Z.

Router(config)#hostname　Router1

Router1(config)#interface　f0/0

Router1(config-if)#ip address 192.168.10.1　255.255.255.0

Router1(config-if)#no　shutdown

Router1(config-if)#exit

Router1(config)#interface   s0/0/0

Router1(config-if)#ip address   12.1.1.1   255.255.255.252

Router1(config-if)#clock   rate   64000

Router1(config-if)#no   shutdown

Router1(config-if)#exit

Router1(config)#

（2）配置 Router2。

Router>enable

Router#configure terminal

Enter configuration commands, one per line.   End with CNTL/Z.

Router(config)#hostname   Router2

Router2(config)#interface   f0/0

Router2(config-if)#ip   address   192.168.20.1   255.255.255.0

Router2(config-if)#no   shutdown

Router2(config-if)#exit

Router2(config)#interface   s0/0/1

Router2(config-if)#ip   address   23.1.1.1   255.255.255.252

Router2(config-if)#no   shutdown

Router2(config-if)#exit

Router2(config)#

（3）配置 Router3（模拟 Internet）。

Router>enable

Router#configure   terminal

Enter configuration commands, one per line.   End with CNTL/Z.

Router(config)#hostname   Router3

Router3(config)#interface   s0/0/0

Router3(config-if)#ip   address   12.1.1.2   255.255.255.252

Router3(config-if)#no   shutdown

Router3(config-if)#exit

Router3(config)#interface   s0/0/1

Router3(config-if)#ip   address   23.1.1.2   255.255.255.252

Router3(config-if)#clock   rate 64000

Router3(config-if)#no   shutdown

Router3(config-if)#exit

Router3(config)#

步骤 2 配置默认路由。

（1）配置 Router1。

Router1(config)#ip route 0.0.0.0 0.0.0.0 12.1.1.2

（2）配置 Router2。

Router2(config)#ip route 0.0.0.0 0.0.0.0 23.1.1.2

（3）测试 PC1 和 PC2 的连通性。

PC>ping 192.168.20.20

Pinging 192.168.20.20 with 32 bytes of data:

Request timed out.

Request timed out.

Request timed out.

Request timed out.

Ping statistics for 192.168.20.20:

    Packets: Sent = 4, Received = 0, Lost = 4 (100% loss),

PC1 发送的数据包经 Router1 的默认路由转发到 Router3，但 Router3 上只有两条直连路由，没有目的网段 192.168.20.0/24 的路由条目，故而丢弃数据包。

步骤 3 配置 GRE VPN。

（1）配置 Router1。

Router1(config)#interface tunnel 1 //创建 Tunnel 接口

Router1(config-if)#ip address 100.1.1.1 255.255.255.0 //IP 地址和 Router2 的隧道接口 IP 应在同一网段

Router1(config-if)#tunnel source s0/0/0 //GRE 封装接口 s0/0/0 的 IP 为公网源 IP

Router1(config-if)#tunnel destination 23.1.1.1 // GRE 封装的公网目标 IP 是 Router2 接口 s0/0/1 的 IP

Router1(config-if)#exit

Router1(config)#ip route 192.168.20.0 255.255.255.0 100.1.1.2 //通过 GRE 到达 192.168.20.0/24 的静态路由下一跳为 Router2 的 tunnel 1 的 IP

（2）配置 Router2。

Router2(config)#interface tunnel 1 //创建 Tunnel 接口

Router2(config-if)#ip address 100.1.1.2 255.255.255.0 //IP 地址和 Router1 的隧道接口 IP 应在同一网段

Router2(config-if)#tunnel source s0/0/1 //GRE 封装接口 s0/0/1 的 IP 为公网源 IP

Router2(config-if)#tunnel destination 12.1.1.1 // GRE 封装的公网目标 IP 是 Router1 接口 s0/0/0 的 IP

Router2(config-if)#exit

Router2(config)#ip route 192.168.10.0 255.255.255.0 100.1.1.1 //通过 GRE 到达

192.168.10.0/24 的静态路由下一跳为 Router1 的 tunnel 1 的 IP

步骤 4 查看路由表。

（1）Router1 的路由表。

Router1#show ip route

Codes: C - connected, S - static, I - IGRP, R - RIP, M - mobile, B - BGP
       D - EIGRP, EX - EIGRP external, O - OSPF, IA - OSPF inter area
       N1 - OSPF NSSA external type 1, N2 - OSPF NSSA external type 2
       E1 - OSPF external type 1, E2 - OSPF external type 2, E - EGP
       i - IS-IS, L1 - IS-IS level-1, L2 - IS-IS level-2, ia - IS-IS inter area
       * - candidate default, U - per-user static route, o - ODR
       P - periodic downloaded static route

Gateway of last resort is 12.1.1.2 to network 0.0.0.0

     12.0.0.0/30 is subnetted, 1 subnets
C   12.1.1.0 is directly connected, Serial0/0/0
     100.0.0.0/24 is subnetted, 1 subnets
C   100.1.1.0 is directly connected, Tunnel1
C   192.168.10.0/24 is directly connected, FastEthernet0/0
S   192.168.20.0/24 [1/0] via 100.1.1.2
S* 0.0.0.0/0 [1/0] via 12.1.1.2

（2）Router2 的路由表。

Router2#show ip route

Codes: C - connected, S - static, I - IGRP, R - RIP, M - mobile, B - BGP
       D - EIGRP, EX - EIGRP external, O - OSPF, IA - OSPF inter area
       N1 - OSPF NSSA external type 1, N2 - OSPF NSSA external type 2
       E1 - OSPF external type 1, E2 - OSPF external type 2, E - EGP
       i - IS-IS, L1 - IS-IS level-1, L2 - IS-IS level-2, ia - IS-IS inter area
       * - candidate default, U - per-user static route, o - ODR
       P - periodic downloaded static route

Gateway of last resort is 23.1.1.2 to network 0.0.0.0

     23.0.0.0/30 is subnetted, 1 subnets
C   23.1.1.0 is directly connected, Serial0/0/1
     100.0.0.0/24 is subnetted, 1 subnets

C   100.1.1.0 is directly connected, Tunnel1
S   192.168.10.0/24 [1/0] via 100.1.1.1
C   192.168.20.0/24 is directly connected, FastEthernet0/0
S*   0.0.0.0/0 [1/0] via 23.1.1.2

（3）Router3 的路由表。

Router3#show   ip   route
Codes: C - connected, S - static, I - IGRP, R - RIP, M - mobile, B - BGP
       D - EIGRP, EX - EIGRP external, O - OSPF, IA - OSPF inter area
       N1 - OSPF NSSA external type 1, N2 - OSPF NSSA external type 2
       E1 - OSPF external type 1, E2 - OSPF external type 2, E - EGP
       i - IS-IS, L1 - IS-IS level-1, L2 - IS-IS level-2, ia - IS-IS inter area
       * - candidate default, U - per-user static route, o - ODR
       P - periodic downloaded static route

Gateway of last resort is not set

     12.0.0.0/30 is subnetted, 1 subnets
C   12.1.1.0 is directly connected, Serial0/0/0
     23.0.0.0/30 is subnetted, 1 subnets
C   23.1.1.0 is directly connected, Serial0/0/1

步骤 5 测试 PC1 和 PC2 的连通性。

PC>ping   192.168.20.20

Pinging 192.168.20.20 with 32 bytes of data:

Request timed out.
Reply from 192.168.20.20: bytes=32 time=6ms TTL=126
Reply from 192.168.20.20: bytes=32 time=5ms TTL=126
Reply from 192.168.20.20: bytes=32 time=2ms TTL=126

Ping statistics for 192.168.20.20:
    Packets: Sent = 4, Received = 3, Lost = 1 (25% loss),
Approximate round trip times in milli-seconds:
    Minimum = 2ms, Maximum = 6ms, Average = 4ms

我们在 Packet Tracer 上抓包，路由器 Router1 接收到 PC1 发送的到达 192.168.20.0/24 的数据包，检查路由表，发现到目的网段数据需要进行 GRE 封装，封装的源 IP 为接口 s0/0/0 的 IP（12.1.1.1），目的 IP 为 Router2 接口 s0/0/1 的 IP（23.1.1.1），封装成 IP 包。然后根据 GRE 封装后的目的 IP 检查路由表，根据默认路由转发到公网。路由器 Router3 上有目的网段 23.1.1.0/30 的直连路由，所以能传送到 Router2 上。Router2 再去除 GRE 包头，根据私网 IP192.168.20.0/24 查找路由表，发现是直连网段，直接转发到目的主机 PC2。如图 6-5-2 和图 6-5-3 所示。

图 6-5-2　GRE 封装

项目 6　连入 Internet

```
设备 Router2 上的PDU信息

OSI模型 | 输入PDU详情 | 输出PDU详情

当前设备: Router2
来源设备: PC0
目的设备: PC1

进入层                                          输出层
应用层                                          应用层
表示层                                          表示层
会话层                                          会话层
传输层                                          传输层
网络层: IP 报头 来源IP: 12.1.1.1,              网络层: IP 报头 来源IP:
目的IP: 23.1.1.1 ICMP Message                   192.168.10.10,目的IP:
类型: 8                                         192.168.20.20 ICMP Message 类
                                                型: 8
数据链路层: HDLC 帧 HDLC                        数据链路层:
物理层: 端口 Serial0/0/1                        物理层:
```

1. The destination IP address matches the IP address of one of the interfaces. The router dispatches the packet to the upper layer.
2. The GRE header of the packet is removed and sent to port Tunnel1.
3. The packet received on Tunnel1 sent up to upper OSI layers.
4. The router looks up the destination IP address in the routing table.

图 6-5-3　去除 GRE 封装

任务 6.6　配置 IPSEC OVER GRE

【实训目的】

配置 IPSEC OVER GRE，实现安全的站点到站点的远程连接。

【实训任务】

1. 配置 GRE VPN。
2. 配置 OSPF。
3. 配置 IPSEC OVER GRE。

【预备知识】

GRE 是一个标准协议，支持多种协议和多播，并支持多点隧道。但是，GRE 缺乏加密机制，数据以明文方式在网络中传输，安全性不足。GRE 可以结合 IPSEC 技术进行加密，GRE OVER IPSEC 和 IPSEC OVER GRE，二者的区别如表 6-6-1 所示。

表 6-6-1

	GRE OVER IPSEC	IPSEC OVER GRE
ACL 定义保护的数据流	GRE 数据流	内网数据流
对端 IP	对端公网地址	对端 GRE Tunnel 地址
应用端口	公网出口	GRE Tunnel

在 Packet Tracer 模拟软件中，我们可以进行 GRE OVER IPSEC 的实验，而 IPSEC OVER GRE 只能在真实的设备上进行实训。

IPSEC OVER GRE 的配置要点如下。

1. IPSECOVER GRE 首先保证 gre 隧道畅通，再让 IPSEC 保护的数据在 GRE 上安全传输。
2. 在通告 OSPF 网段时，需要通告内网各网段和 GRE Tunnel 网段，切忌通告公网出口网段。
3. IPSECOVERGRE 先进行 IPSEC 封装，再进行 GRE 封装。所以定义 ACL 时保护的数据流为内网数据流。
4. 路由器品牌不同，定义 IPSEC 的远端地址不同。锐捷路由器为 GRE tunnel 的 IP，思科路由器为对方公网出口的 IP。

【实训拓扑】

网络拓扑结构图如图 6-6 所示。

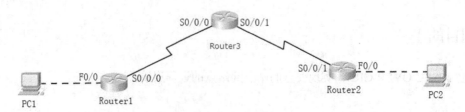

图 6-6　网络拓扑结构图

【实训设备】

路由器 3 台、计算机 2 台。

【实训步骤】

步骤 1 按照拓扑图搭建网络,并配置各设备 IP 地址,参见表 6-6-2 所示。

表 6-6-2

设备名称	接口	IP	子网掩码	网关
Router1	S0/0/0	12.1.1.1	255.255.255.252	
	F0/0	192.168.10.1	255.255.255.0	
Router2	S0/0/1	23.1.1.1	255.255.255.252	
	F0/0	192.168.20.1	255.255.255.0	
Router3	S0/0/0	12.1.1.2	255.255.255.252	
	S0/0/0	23.1.1.2	255.255.255.252	
PC1	网卡	192.168.10.10	255.255.255.0	192.168.10.1
PC2	网卡	192.168.20.20	255.255.255.0	192.168.20.1

(1)配置 Router1。

Router>enable

Router#configure Terminal

Enter configuration commands, one per line.   End with CNTL/Z.

Router(config)#hostname   Router1

Router1(config)#interface   f0/0

Router1(config-if)#ip address 192.168.10.1   255.255.255.0

Router1(config-if)#no   shutdown

Router1(config-if)#exit

Router1(config)#interface   s0/0/0

Router1(config-if)#ip address   12.1.1.1   255.255.255.252

Router1(config-if)#clock   rate   64000

Router1(config-if)#no   shutdown

Router1(config-if)#exit

Router1(config)#

(2)配置 Router2。

Router>enable

Router#configure terminal

Enter configuration commands, one per line. End with CNTL/Z.

Router(config)#hostname Router2

Router2(config)#interface f0/0

Router2(config-if)#ip address 192.168.20.1 255.255.255.0

Router2(config-if)#no shutdown

Router2(config-if)#exit

Router2(config)#interface s0/0/1

Router2(config-if)#ip address 23.1.1.1 255.255.255.252

Router2(config-if)#no shutdown

Router2(config-if)#exit

Router2(config)#

（3）配置 Router3（模拟 Internet）。

Router>enable

Router#configure terminal

Enter configuration commands, one per line. End with CNTL/Z.

Router(config)#hostname Router3

Router3(config)#interface s0/0/0

Router3(config-if)#ip address 12.1.1.2 255.255.255.252

Router3(config-if)#no shutdown

Router3(config-if)#exit

Router3(config)#interface s0/0/1

Router3(config-if)#ip address 23.1.1.2 255.255.255.252

Router3(config-if)#clock rate 64000

Router3(config-if)#no shutdown

Router3(config-if)#exit

Router3(config)#

步骤 2 配置默认路由。

（1）配置 Router1。

Router1(config)#ip route 0.0.0.0 0.0.0.0 12.1.1.2

（2）配置 Router2。

Router2(config)#ip route 0.0.0.0 0.0.0.0 23.1.1.2

（3）测试 PC1 和 PC2 的连通性。

PC>ping 192.168.20.20

Pinging 192.168.20.20 with 32 bytes of data:

Request timed out.

Request timed out.

Request timed out.

Request timed out.

Ping statistics for 192.168.20.20:

Packets: Sent = 4, Received = 0, Lost = 4 (100% loss),

PC1 发送的数据包经 Router1 的默认路由转发到 Router3，但 Router3 上只有两条直连路由，没有目的网段 192.168.20.0/24 的路由条目，故而丢弃数据包。

步骤 3 配置 GRE VPN。

（1）配置 Router1。

Router1(config)#interface tunnel 1 //创建 Tunnel 接口

Router1(config-if)#ip address 100.1.1.1 255.255.255.0 //IP 地址和 Router2 的隧道接口 IP 应在同一网段

Router1(config-if)#tunnel source s0/0/0 //GRE 封装接口 s0/0/0 的 IP 为公网源 IP

Router1(config-if)#tunnel destination 23.1.1.1 // GRE 封装的公网目标 IP 是 Router2 接口 s0/0/1 的 IP

Router1(config-if)#exit

（2）配置 Router2。

Router2(config)#interface tunnel 1 //创建 Tunnel 接口

Router2(config-if)#ip address 100.1.1.2 255.255.255.0 //IP 地址和 Router1 的隧道接口 IP 应在同一网段

Router2(config-if)#tunnel source s0/0/1 //GRE 封装接口 s0/0/1 的 IP 为公网源 IP

Router2(config-if)#tunnel destination 12.1.1.1 // GRE 封装的公网目标 IP 是 Router1 接口 s0/0/0 的 IP

Router2(config-if)#exit

步骤 4 配置 OSPF。

（1）配置 Router1。

Router1(config)#router ospf 1

Router1(config-router)#router-id 1.1.1.1

Router1(config-router)#network 192.168.10.0 0.0.0.255 area 0

Router1(config-router)#network 100.1.1.0 0.0.0.255 area 0

Router1(config-router)#end

Router1#

（2）配置 Router2。

Router2(config)#router ospf 1

Router2(config-router)#router-id 2.2.2.2

Router2(config-router)#network   192.168.20.0   0.0.0.255   area   0

Router2(config-router)#network   100.1.1.0   0.0.0.255   area   0

Router2(config-router)#end

Router2#

（3）查看路由表。

Router1#show   ip   route

Codes: C - connected, S - static, I - IGRP, R - RIP, M - mobile, B - BGP

　　　　D - EIGRP, EX - EIGRP external, O - OSPF, IA - OSPF inter area

　　　　N1 - OSPF NSSA external type 1, N2 - OSPF NSSA external type 2

　　　　E1 - OSPF external type 1, E2 - OSPF external type 2, E - EGP

　　　　i - IS-IS, L1 - IS-IS level-1, L2 - IS-IS level-2, ia - IS-IS inter area

　　　　* - candidate default, U - per-user static route, o - ODR

　　　　P - periodic downloaded static route

Gateway of last resort is 12.1.1.2 to network 0.0.0.0

　　　12.0.0.0/30 is subnetted, 1 subnets

C　　　12.1.1.0 is directly connected, Serial0/0/0

　　　100.0.0.0/24 is subnetted, 1 subnets

C　　　100.1.1.0 is directly connected, Tunnel1

C　　192.168.10.0/24 is directly connected, FastEthernet0/0

O　　192.168.20.0/24 [110/1001] via 100.1.1.2, 00:01:17, Tunnel1

S*　　0.0.0.0/0 [1/0] via 12.1.1.2

Router2#show   ip   route

Codes: C - connected, S - static, I - IGRP, R - RIP, M - mobile, B - BGP

　　　　D - EIGRP, EX - EIGRP external, O - OSPF, IA - OSPF inter area

　　　　N1 - OSPF NSSA external type 1, N2 - OSPF NSSA external type 2

　　　　E1 - OSPF external type 1, E2 - OSPF external type 2, E - EGP

　　　　i - IS-IS, L1 - IS-IS level-1, L2 - IS-IS level-2, ia - IS-IS inter area

　　　　* - candidate default, U - per-user static route, o - ODR

　　　　P - periodic downloaded static route

Gateway of last resort is 23.1.1.2 to network 0.0.0.0

　　　23.0.0.0/30 is subnetted, 1 subnets

C　　　23.1.1.0 is directly connected, Serial0/0/1

100.0.0.0/24 is subnetted, 1 subnets

C   100.1.1.0 is directly connected, Tunnel1

O   192.168.10.0/24 [110/1001] via 100.1.1.1, 00:01:52, Tunnel1

C   192.168.20.0/24 is directly connected, FastEthernet0/0

S*   0.0.0.0/0 [1/0] via 23.1.1.2

以上输出结果显示，Router1 和 Router2 通过 GRE 互相学习到了对方的网段路由条目。

步骤 5　配置 IPSEC VPN。

（1）配置 Router1。

Router1(config)#access-list 110 permit ip 192.168.10.0 0.0.0.255 192.168.20.0 0.0.0.255　　//定义要保护的数据流

Router1(config)#crypto isakmp policy 10

Router1(config-isakmp)#encr 3des

Router1(config-isakmp)#hash md5

Router1(config-isakmp)#authentication pre-share

Router1(config-isakmp)#exit

Router1(config)#crypto isakmp key ruijie address 100.1.1.2 //锐捷路由器对端 IP 为隧道口 tunnel 1 的 IP，思科路由器为对端物理口 s0/0/1 的 IP

Router1(config)#crypto ipsec transform-set test esp-3des esp-md5-hmac

Router1(config)#crypto map test 10 ipsec-isakmp

Route1(config-crypto-map)#set peer 100.1.1.2　　//锐捷路由器对端 IP 为隧道口 tunnel 1 的 IP，思科路由器为对端物理口 s0/0/1 的 IP

Route1(config-crypto-map)#set transform-set test

Route1(config-crypto-map)# match address 110

Route1(config-crypto-map)#exit

Route1(config)#interface   tunnel 1

Router1(config-if)#crypto   map   test

Router1(config-if)#end

（2）配置 Router2。

Router2(config)#access-list 110 permit ip 192.168.20.0 0.0.0.255 192.168.10.0 0.0.0.255

Router2(config)#crypto isakmp policy 10

Router2(config-isakmp)#encr 3des

Router2(config-isakmp)#hash md5

Router2(config-isakmp)#authentication pre-share

Router2(config-isakmp)#exit

Router2(config)#crypto isakmp key ruijie address 100.1.1.1　　//锐捷路由器对端 IP 为隧道口

tunnel 1 的 IP，思科路由器为对端物理口 s0/0/0 的 IP

Router2(config)#crypto ipsec transform-set test esp-3des esp-md5-hmac

Router2(config)#crypto map test 10 ipsec-isakmp

Route2(config-crypto-map)#set peer 100.1.1.1   //锐捷路由器对端 IP 为隧道口 tunnel 1 的 IP，思科路由器为对端物理口 s0/0/0 的 IP

Route2(config-crypto-map)#set transform-set test

Route2(config-crypto-map)# match address 110

Route2(config-crypto-map)#exit

Route2(config)#interface   tunnel 1

Router2(config-if)#crypto   map   test

Router2(config-if)#end

步骤 6 验证测试。

PC>ping   192.168.20.20

Pinging 192.168.20.20 with 32 bytes of data:

Request timed out.
Reply from 192.168.20.20: bytes=32 time=5ms TTL=126
Reply from 192.168.20.20: bytes=32 time=5ms TTL=126
Reply from 192.168.20.20: bytes=32 time=5ms TTL=126

Ping statistics for 192.168.20.20:
    Packets: Sent = 4, Received = 3, Lost = 1 (25% loss),
Approximate round trip times in milli-seconds:
    Minimum = 5ms, Maximum = 5ms, Average = 5ms

项目 7 竞赛试题

本项目精心挑选了 2010 年至 2013 年青岛市和全国技能大赛中的四套企业网搭建及应用试题。通过本项目的练习，可以开阔视野，增强综合运用所学知识的能力，并且使学生对各级技能大赛有一个初步的认识。

2010 年青岛市中等职业学校企业网搭建与应用技能比赛

第一场试题

设备准备

1. 服务器（Server）两台（注：没有安装操作系统）
2. 工作站（PC） 三台（注：没有安装操作系统）
3. 路由器（RG-RSR20）两台 + 交换机三台(RG-S3760-24 两台,RG-S2126S 一台)
4. RJ-45 钳子   两把
5. RJ-45 水晶头   若干个
6. 网线测试仪   一个
7. 双绞线   若干米
8. Windows 2003 Server 安装光盘
9. Windows XP 安装光盘
10. Linux 安装光盘

安装与调试步骤如下。

按照国际标准（568A/568B）制作实验所需数量的双绞线并保证其连通。组成一个小型局域网。连接示意图如图 7-1 所示。

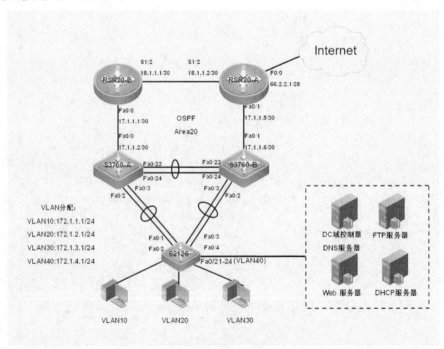

图 7-1 网络拓扑结构图

地址规划列表如表 7-1 所示。

表 7-1

源设备名称	设备接口	IP 地址	备注
RSR20-A	S1/2	18.1.1.2/30	
RSR20-A	FA0/0	66.2.2.1/28	
RSR20-A	FA0/1	17.1.1.5/30	
RSR20-B	FA0/0	17.1.1.1/30	
RSR20-B	S1/2	18.1.1.1/30	
S3760-A	FA0/1	17.1.1.2/30	
S3760-A	VLAN10	172.1.1.1/24	
S3760-A	VLAN20	172.1.2.1/24	
S3760-A	VLAN30	172.1.3.1/24	
S3760-A	VLAN40	172.1.4.1/24	
S3760-B	VLAN10	172.1.1.2/24	
S3760-B	VLAN20	172.1.2.2/24	
S3760-B	VLAN30	172.1.3.2/24	
S3760-B	VLAN40	172.1.4.2/24	
DC/DNS 服务器	PC1	172.1.4.10/24	
FTP 服务器	PC2	172.1.4.11/24	
Web 服务器	PC1	172.1.4.12/24	
DHCP 服务器	PC1	172.1.4.13/24	

竞赛需求

一、系统要求（30 分）

DC/DNS 服务器（5 分）	
虚拟机系统	安装 Windows Server 2003，地址为 172.1.4.10/24
虚拟机名称	ADSERVER
服务配置	活动目录服务：配置为域控制器，FQDN：adserver.Labtest.com
	DNS 服务：配置 DNS，能够正确配置正向、反向解析，能够解析 Web、FTP、DHCP 服务器，并且允许动态更新。
	创建用户 user1、user2、user3、user4

FTP 服务器（10 分）

续表

独立服务器	安装 Linux 系统，地址为 172.1.4.11/24
服务器名称	Nfsserver
服务配置	FQDN:ftpserver.labtest.com 系统默认启动为运行级别 3 创建目录/mnt/share1、/mnt/share 允许所有主机可以挂载 share 目录，但只有读权限 只允许 172.1.4.0/24 网段挂载 share1 目录，可以有读写权限 创建文件的属主和属组的用户 ID 为 511； 要求系统启动时自动挂载

Web 服务器（10 分）

虚拟机系统	安装 Windows Server 2003 系统，IP 地址为 172.1.4.12/24
虚拟机名称	Webserver
服务配置	FQDN：webserver.labtest.com 添加 IIS 组件 创建一个 Web 站点，发布一个"预祝大赛圆满成功"的简单网页信息。 在工作站上输入 www.testnetwork 能够浏览上面发布的简单网页信息。

DHCP 服务器（5 分）

虚拟机系统	安装 Windows Server 2003 系统，IP 地址为 172.1.4.10/24
虚拟机名称	dhcpserver
服务配置	FQDN：dhcpserver.labtest.com 创建作用域和超级作用域，为 VLAN10、VLAN20、VLAN30 分配 IP 地址 VLAN10：网关 172.1.1.254、DNS172.1.4.10，排除 172.1.1.1-172.1.1.10 VLAN20：网关 172.1.2.254、DNS172.1.4.10，排除 172.1.2.1-172.1.2.10 VLAN30：网关 172.1.3.254、DNS172.1.4.10，排除 172.1.3.1-172.1.3.10

二、网络要求（70 分）

1. 路由器 RSR20-A 需求（14 分）

基本功能（2 分）	配置各接口 IP 地址
路由功能（3 分）	配置 OSPF 协议、静态路由或路由重分发，指定 route-id 为 3.3.3.3，使全网互通
NAT 功能（5 分）	VLAN10 能够通过地址池（66.2.2.3～66.2.2.4/28）访问互联网 VLAN20 能够通过地址池（66.2.2.5～66.2.2.6/28）访问互联网 VLAN30 能够通过地址池（66.2.2.7～66.2.2.8/28）访问互联网 将 Web 服务器发布到互联网上，其公网 IP 地址为 66.2.2.10
安全功能（4 分）	配置 ACL 实现 VLAN10、VLAN20、VLAN30 用户只有上班时间（周一至周五的 9：00～18：00）才能访问互联网

2. 路由器 RSR20-B 需求（5 分）

基本功能（2 分）	配置各接口 IP 地址
路由功能（3 分）	配置 OSPF 协议、静态路由或路由重分发，指定 route-id 为 4.4.4.4，使全网互通

3. 交换机 S3760-A 需求（20 分）

基本功能（2 分）	配置各接口 IP 地址、创建并配置 VLAN
路由功能（5 分）	配置 OSPF 路由协议，指定 route-id 为 2.2.2.2 使全网互通
优化功能（5 分）	配置 MSTP 协议 创建两个 MSTP 实例：Instance10、Instance20 Instance10 包括：VLAN10、VLAN20 Instance20 包括：VLAN30、VLAN40 此交换机为 Instance10 的生成树根，是 Instance20 的生成树备份根 配置 DHCP 中继
可靠性能（5 分）	配置 VRRP 协议，创建四个 VRRP 组，分别为 group10、group20、group30、group40 S3760-A 是 VLAN10、VLAN20 的活跃路由器，VLAN30、VLAN40 的备份路由器 配置链路聚合，将 FA0/23-24 两个接口配置为链路聚合，并将聚合口配置为 TRUNK。将 FA0/2-3 两个接口配置为链路聚合，并将聚合口配置为 TRUNK
安全功能（3 分）	不允许 VLAN10 与 VLAN20 互相访问，其他不受限制

4. 交换机 S3760-B 需求（18 分）

基本功能（2 分）	配置各接口 IP 地址、创建并配置 VLAN
路由功能（3 分）	配置 OSPF 路由协议，指定 route-id 为 1.1.1.1 使全网互通
优化功能（5 分）	配置 MSTP 协议 创建两个 MSTP 实例：Instance10、Instance20 Instance10 包括：VLAN10、VLAN20 Instance20 包括：VLAN30、VLAN40 此交换机为 Instance20 的生成树根，是 Instance10 的生成树备份根 配置 DHCP 中继
可靠性能（5 分）	配置 VRRP 协议，创建四个 VRRP 组，分别为 group10、group20、group30、group40 S3760-B 是 VLAN30、VLAN40 的活跃路由器，VLAN10、VLAN20 的备份路由器 配置链路聚合，将 FA0/23-24 两个接口配置为链路聚合，并将聚合口配置为 TRUNK。将 FA0/2-3 两个接口配置为链路聚合，并将聚合口配置为 TRUNK
安全功能（3 分）	不允许 VLAN10 与 VLAN20 互相访问，其他不受限制

5. 交换机 S2126 需求（13 分）

基本功能（2 分）	创建并配置 VLAN
优化功能（3 分）	配置 MSTP 协议 创建两个 MSTP 实例：Instance10、Instance20 Instance10 包括：VLAN10 Instance20 包括：VLAN20
可靠性能（4 分）	配置链路聚合，将 FA0/1-2 两个接口配置为链路聚合，并将聚合口配置为 TRUNK。将 FA0/3-4 两个接口配置为链路聚合，并将聚合口配置为 TRUNK
安全功能（4 分）	配置端口安全功能，每个接入接口的最大连接数为 2，如果违规则关闭端口

三、参考答案

[S2126]

Switch>enable

Switch#configure terminal

Enter configuration commands, one per line.　End with CNTL/Z.

Switch(config)#hostname S2126

S2126(config)#vlan range 10,20,30,40

S2126(config-vlan-range)#exit

S2126(config)#interface range fastEthernet 0/21-24

S2126(config-if-range)#switchport access vlan 40

S2126(config)#spanning-tree

S2126(config)#spanning-tree mst configuration

S2126(config-mst)#instance 10 vlan 10,20

S2126(config-mst)#instance 20 vlan 30,40

S2126(config)#interface range fastEthernet 0/1-2

S2126(config-if-range)#port-group 2

S2126(config)#int aggregatePort 2

S2126(config-if)#switchport mode trunk

S2126(config)#interface range fastEthernet 0/3-4

S2126(config-if-range)#port-group 3

S2126(config)#int aggregatePort 3

S2126(config-if)#switchport mode trunk

S2126(config)#interface range fastEthernet 0/5-24

S2126(config-if-range)#switchport port-security

S2126(config-if-range)#switchport port-security maximum 2

S2126(config-if-range)#switchport port-security violation shutdown

[S3760-A]

Ruijie>enable
Ruijie#configure terminal
Ruijie(config)#hostname S3760-A
S3760-A(config)#service dhcp
S3760-A(config)#ip helper-address 172.1.4.13
S3760-A(config)#
S3760-A(config)#vlan range 10,20,30,40
S3760-A(config-vlan-range)#exit
S3760-A(config)#int range fastEthernet 0/23-24
S3760-A(config-if-range)#port-group 1
S3760-A(config-if-range)#exit
S3760-A(config)#interface aggregateport 1
S3760-A(config-AggregatePort 1)#switchport mode trunk
S3760-A(config)#int range fastEthernet 0/2-3
S3760-A(config-if-range)#port-group 2
S3760-A(config)#interface aggregateport 2
S3760-A(config-AggregatePort 2)#switchport mode trunk
S3760-A(config)#interface fastEthernet 0/1
S3760-A(config-FastEthernet 0/1)#no switchport
S3760-A(config-FastEthernet 0/1)#ip address 17.1.1.2 255.255.255.252
S3760-A(config)#interface vlan 10
S3760-A(config-VLAN 10)#ip add 172.16.1.1 255.255.255.0
S3760-A(config-VLAN 10)#vrrp 10 ip 172.1.1.254
S3760-A(config-VLAN 10)#vrrp 10 priority 120
S3760-A(config-VLAN 10)#ip access-group 110 in
S3760-A(config)#interface vlan 20
S3760-A(config-VLAN 20)#ip add 172.16.2.1 255.255.255.0
S3760-A(config-VLAN 20)#vrrp 20 ip 172.1.2.254
S3760-A(config-VLAN 20)#vrrp 20 priority 120
S3760-A(config)#interface vlan 30
S3760-A(config-VLAN 30)#ip add 172.16.3.1 255.255.255.0
S3760-A(config-VLAN 30)#vrrp 30 ip 172.1.3.254

S3760-A(config)#interface vlan 40

S3760-A(config-VLAN 40)#ip add 172.16.4.1 255.255.255.0

S3760-A(config-VLAN 40)#vrrp 40 ip 172.1.4.254

S3760-A(config-VLAN 40)#exit

S3760-A(config)#router ospf 1

S3760-A(config-router)#router-id 2.2.2.2

Change router-id and update OSPF process! [yes/no]:y

S3760-A(config-router)#network 17.1.1.0 0.0.0.3 a 20

S3760-A(config-router)#network 172.16.1.0 0.0.0.255 a 20

S3760-A(config-router)#network 172.16.2.0 0.0.0.255 a 20

S3760-A(config-router)#network 172.16.3.0 0.0.0.255 a 20

S3760-A(config-router)#network 172.16.4.0 0.0.0.255 a 20

S3760-A(config)#spanning-tree

S3760-A(config)#spanning-tree mst configuration

S3760-A(config-mst)#instance 10 vlan 10,20

S3760-A(config-mst)#instance 20 vlan 30,40

S3760-A(config)#spanning-tree mst 10 priority 4096

S3760-A(config)#spanning-tree mst 20 priority 8192

S3760-A(config)#access-list 110 deny ip 172.1.1.0 0.0.0.255 172.1.2.0 0.0.0.255

S3760-A(config)#access-list 110 permit ip any any

[S3760-B]

Ruijie>enable

Ruijie#configure terminal

Ruijie(config)#hostname S3760-B

S3760-B(config)#service dhcp

S3760-B(config)#ip helper-address 172.1.4.13

S3760-B(config)#vlan range 10,20,30,40

S3760-B(config-vlan-range)#exit

S3760-B(config)#int range fastEthernet 0/23-24

S3760-B(config-if-range)#port-group 1

S3760-B(config-if-range)#exit

S3760-B(config)#interface aggregateport 1

S3760-B(config-AggregatePort 1)#switchport mode trunk

S3760-B(config)#int range fastEthernet 0/2-3

S3760-B(config-if-range)#port-group 3
S3760-B(config)#interface aggregateport 3
S3760-B(config-AggregatePort 3)#switchport mode trunk
S3760-B(config)#interface fastEthernet 0/1
S3760-B(config-FastEthernet 0/1)#no switchport
S3760-B(config-FastEthernet 0/1)#ip address 17.1.1.6 255.255.255.252
S3760-B(config)#interface vlan 10
S3760-B(config-VLAN 10)#ip add 172.16.1.2 255.255.255.0
S3760-B(config-VLAN 10)#vrrp 10 ip 172.1.1.254
S3760-B(config-VLAN 10)#ip access-group 110 in
S3760-B(config)#interface vlan 20
S3760-B(config-VLAN 20)#ip add 172.16.2.2 255.255.255.0
S3760-B(config-VLAN 20)#vrrp 20 ip 172.1.2.254
S3760-B(config)#interface vlan 30
S3760-B(config-VLAN 30)#ip add 172.16.3.2 255.255.255.0
S3760-B(config-VLAN 30)#vrrp 30 ip 172.1.3.254
S3760-B(config-VLAN 30)#vrrp 30 priority 120
S3760-B(config)#interface vlan 40
S3760-B(config-VLAN 40)#ip add 172.16.4.2 255.255.255.0
S3760-B(config-VLAN 40)#vrrp 40 ip 172.1.4.254
S3760-B(config-VLAN 40)#vrrp 40 priority 120
S3760-B(config-VLAN 40)#exit
S3760-B(config)#router ospf 1
S3760-B(config-router)#router-id 1.1.1.1
Change router-id and update OSPF process! [yes/no]:y
S3760-B(config-router)#network 17.1.1.40.0.0.3 a 20
S3760-B(config-router)#network 172.16.1.00.0.0.255 a 20
S3760-B(config-router)#network 172.16.2.00.0.0.255 a 20
S3760-B(config-router)#network 172.16.3.00.0.0.255 a 20
S3760-B(config-router)#network 172.16.4.00.0.0.255 a 20
S3760-B(config)#spanning-tree
S3760-B(config)#spanning-tree mst configuration
S3760-B(config-mst)#instance 10 vlan 10,20
S3760-B(config-mst)#instance 20 vlan 30,40
S3760-B(config)#spanning-tree mst 20 priority 4096
S3760-B(config)#spanning-tree mst 10 priority 8192

S3760-B(config)#access-list 110 deny ip 172.1.1.0 0.0.0.255 172.1.2.0 0.0.0.255

S3760-B(config)#access-list 110 permit ip any any

[RSR20-A]

Ruijie>enable

Ruijie#configure terminal

Ruijie(config)#hostname RSR20-A

RSR20-A(config)#interface fastEthernet 0/1

RSR20-A(config-if-FastEthernet 0/1)#ip nat inside

RSR20-A(config-if-FastEthernet 0/1)#ip address 17.1.1.5 255.255.255.252

RSR20-A(config-if-FastEthernet 0/1)#exit

RSR20-A(config)#interface serial 2/0

RSR20-A(config-if-Serial 2/0)#ip add 18.1.1.2 255.255.255.252

RSR20-A(config-if-Serial 2/0)#clock rate 128000

RSR20-A(config-if-Serial 2/0)#ip nat inside

RSR20-A(config)#interface fastEthernet 0/0

RSR20-A(config-if-FastEthernet 0/0)#ip address 66.2.2.1 255.255.255.240

RSR20-A(config-if-FastEthernet 0/0)#ip nat outside

RSR20-A(config)#ip route 0.0.0.0 0.0.0.0 fastEthernet 0/0

RSR20-A(config)#router ospf 1

RSR20-A(config-router)#router-id 3.3.3.3

Change router-id and update OSPF process! [yes/no]:y

RSR20-A(config-router)#network 17.1.1.4 0.0.0.3 a 20

RSR20-A(config-router)#network 18.1.1.0 0.0.0.3 a 20

RSR20-A(config-router)#default-information originate always

RSR20-A(config-router)#exit

RSR20-A(config)#time-range time

RSR20-A(config-time-range)#periodic weekdays 9:00 to 18:00

RSR20-A(config-time-range)#exit

RSR20-A(config)#access-list 10 permit 172.16.1.0 0.0.0.255 time-range time

RSR20-A(config)#access-list 20 permit 172.16.2.0 0.0.0.255 time-range time

RSR20-A(config)#access-list 30 permit 172.16.3.0 0.0.0.255 time-range time

RSR20-A(config)#ip nat pool vlan10   66.2.2.3   66.2.2.4   netmask   255.255.255.240

RSR20-A(config)#ip nat pool vlan20   66.2.2.5   66.2.2.6   netmask   255.255.255.240

RSR20-A(config)#ip nat pool vlan30   66.2.2.7   66.2.2.8   netmask   255.255.255.240

RSR20-A(config)#ip nat inside source list 10 pool 10 overload
RSR20-A(config)#ip nat inside source list 20 pool 20 overload
RSR20-A(config)#ip nat inside source list 30 pool 30 overload
RSR20-A(config)# ip nat inside source static tcp 172.16.4.12 80 66.2.2.10 80 permit-inside

[RSR20-B]

Ruijie>enable
Ruijie#configure terminal
Ruijie(config)#hostname RSR20-B
RSR20-B(config)#interface fastEthernet 0/0
RSR20-B(config-if-FastEthernet 0/0)#ip address 17.1.1.1 255.255.255.252
RSR20-B(config-if-FastEthernet 0/0)#exit
RSR20-B(config)#interface serial 2/0
RSR20-B(config-if-Serial 2/0)#ip address 18.1.1.1 255.255.255.252
RSR20-B(config-if-Serial 2/0)#exit
RSR20-B(config)#router ospf 1
RSR20-B(config-router)#router-id 4.4.4.4
Change router-id and update OSPF process! [yes/no]:y
RSR20-B(config-router)#network 17.1.1.0 0.0.0.3 a 20
RSR20-B(config-router)#network 18.1.1.0 0.0.0.3 a 20

2011 年全国中职技能大赛企业网搭建及应用模拟试题

一、注意事项

1. 检查硬件设备、网线头、Console 线等物品的数量是否齐全，电脑设备是否正常。
2. 自带双绞线制作和验证测试工具。禁止携带和使用移动存储设备、运算器、通信工具及参考资料。
3. 操作完成后，需要保存设备配置，不要关闭任何设备，不要拆动硬件的连接，不要对设备随意加设密码，试卷留在考场。
4. 请勿损坏赛场准备的比赛所需要竞赛设备、竞赛软件和竞赛材料等。

二、竞赛环境

硬件环境：

设备类型	设备型号	设备数量（台）
路由器	RG-RSR20-18	4
三层交换机	RG-S3760E-24	2
无线 AP	RG-AP220E	1
计算机	—	5

软件环境：

软件名称	介质形式	软件数量
Windows 2003 Server 企业版	光盘	1
Windows XP Professional	光盘	1

三、网络拓扑

某集团在全国有两家公司，即总公司和分公司。总公司电脑数量较多，分公司电脑数量较少。总公司和分公司通过穿越 internet 建立 vpn 隧道来进行内部访问。同时总公司和分公司的员工都可以访问互联网。分公司内部含有一台运行着 Web 服务和 FTP 服务的服务器，需要对外提供服务。如果你是这个网络项目的网络工程师，请根据下面的需求构建一个安全、稳定的网络。

已知总公司有两个大部门，需要创建 VLAN 10 和 VLAN 20，每个部门中有 100 台主机。HOST1 是 VLAN 10 中的一台，HOST2 是 VLAN 20 中的一台。上级领导只给了 192.168.1.0/24 的网段。需要自己划分出 VLAN 10 所在的网段和 VLAN 20 所在的网段。

分公司的员工较少，均采用无线连入办公网络。自动获取 192.168.3.8/29 这个网段的 IP 地址来进行通信。如图 7-2 所示。

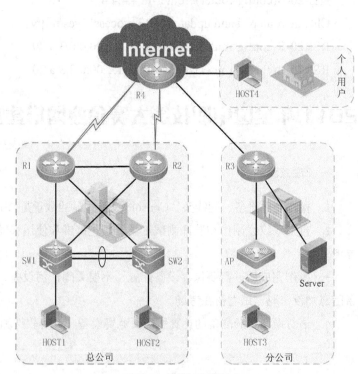

图 7-2　网络拓扑结构图

四、IP 规划（如表 7-2 所示）

表 7-2

地理位置	设备	端口	IP	对端设备	接口	备注
总公司	R1	F0/0	192.168.2.1/30	R2	F0/0	
		F0/1	192.168.2.9/29	SW1	F0/11	
		F0/2	192.168.2.17/29	SW2	F0/11	
		S2/0	200.1.1.1/30	R4	S2/0	
	R2	F0/0	192.168.2.2/30	R1	F0/0	
		F0/1	192.168.2.10/29	SW1	F0/12	
		F0/2	192.168.2.18/29	SW2	F0/12	
		S2/0	200.1.1.5/30	R4	S2/1	
	SW1	F0/1-5	自动获取	HOST1	NIC	
		F0/11	Vlan 30	R1	F0/1	
		F0/12	Vlan 30	R2	F0/1	
		F0/13-16	聚合	SW2	F0/13-16	
		VLAN 10	192.168.1.126			
		VLAN 20	192.168.1.254			
		VLAN 30	192.168.2.11			
	SW2	F0/6-10	自动获取	HOST2	NIC	
		F0/11	Vlan 40	R1	F0/2	
		F0/12	Vlan 40	R2	F0/2	
		F0/13-16	聚合	SW1	F0/13-16	
		VLAN 10	192.168.1.125			
		VLAN 20	192.168.1.253			
		VLAN 40	192.168.2.19			
	HOST1	NIC	自动获取	SW1	F0/1-5	
	HOST2	NIC	自动获取	SW2	F0/6-10	
分公司	R3	F0/0	200.1.1.9/30	R4	F0/0	
		F0/1	192.168.3.1/30	AP	G0	
		F0/2	192.168.3.5/30	SERVER	NIC	
	AP	G0	192.168.3.2/30	R3	F0/1	
		WLAN	192.168.3.9/29	HOST	WLAN	
	SERVER	NIC	192.168.3.6/30	R3	F0/2	
	HOST3	WLAN	自动获取	AP	WLAN	
个人用户	R4	S2/0	200.1.1.2/30	R1	S2/0	
		S2/1	200.1.1.6/30	R2	S2/1	
		F0/0	200.1.1.10/30	R3	F0/0	
		F0/1	200.1.1.14	HOST4	NIC	
	HOST4	NIC	200.1.1.13	R4	F0/1	

五、试题内容

（一）网络系统构建(95分)

（1）网络底层配置。（24分）

如图7-2所示连接所有线缆。（2分）

配置SW1的1-5口连接VLAN 10，HOST1在VLAN 10中。（2分）

配置SW1的6-10口连接VLAN 20。（2分）

配置SW2的1-5口连接VLAN 10。（2分）

配置SW2的6-10口连接VLAN 20，HOST2在VLAN 20中。（2分）

为SW1和SW2的VLAN配置IP地址。（2分）

如表7-2所示，创建聚合。（2分）

配置连接主机的端口最大连接数为4，违例状态为关闭。（2分）

配置DHCP服务，使HOST1和HOST2均能够获得正确的IP地址。（8分）

（2）路由协议配置。（16分）

分别在R1、R2、SW1、SW2配置动态路由协议OSPF。（4分）

在R1、R2上配置默认路由，并重新分发到OSPF中。（4分）

分别在R1和R2上配置VRRP组10和20。用于默认将SW1提交的数据由R1转发，默认将SW2提交的数据由R2转发。（8分）

（3）网络出口配置。（18分）

分别在R1和R2上配置NAT，使用内部VLAN10、VLAN20的用户可以使用外部接口的IP地址访问互联网。（6分）

分别在R1和R2上配置访问控制列表，允许内部的员工使用域名访问外网带SSL的链接的Web服务。（6分）

分别在R1和R2上配置访问控制列表，用于禁止外网访问内网的138和139端口。（6分）

（4）公司间互访。（10分）

在R2上建立总公司和分公司之间的VPN互联通道。使总公司和分公司的主机及服务器之间可以通过VPN进行互访。（5分）

在R3上建立总公司和分公司之间的VPN互联通道。使总公司和分公司的主机及服务器之间可以通过VPN进行互访。（5分）

（5）分公司配置。（19分）

在R3上配置NAT，使内部的无线员工能够使用域名访问外网带SSL的链接的Web服务。（5分）

在R3上配置静态NAT，将服务器的Web服务和FTP服务映射到公网上。（5分）

配置无线AP，采用WEP加密方式，40位ASCII密码形式，口令为12345。（3分）

配置无线AP，使其SSID为Ruijie，不允许SSID广播。（3分）

配置无线AP的DHCP服务，使通过无线连接上来的客户端都能够获得正确的IP。（3分）

（6）其他配置。（8分）

如表 7-1 所示，配置 HOST4 的 IP 地址。（2分）

如表 7-1 和图 7-2 所示，配置 R4 的各接口 IP 地址。（2分）

在 SERVER 上搭建 Web 服务和 FTP 服务。（4分）

（二）整体实现及测试（5分）

HOST1、HOST2、HOST4 可以访问 SERVER 的 Web 和 FTP 服务。（3分）

HOST1、HOST2、HOST3 可以互访。（2分）

六、参考答案

[SW1]

Ruijie>enable

Ruijie#configure terminal

Enter configuration commands, one per line. End with CNTL/Z.

Ruijie(config)#hostname SW1

SW1(config)#

SW1(config)#vlan range 10,20,30,40

SW1(config-vlan-range)#exit

SW1(config)#interface range f0/1-5

SW1(config-if-range)#switchport access vlan 10

SW1(config)#interface range f 0/6-10

SW1(config-if-range)#switchport access vlan 20

SW1(config-if-range)#exit

SW1(config)#interface range f0/13-16

SW1(config-if-range)#port-group 1

SW1(config-if-range)#interface agg 1

SW1(config-AggregatePort 1)#switchport mode trunk

SW1(config)#interface vlan 10

SW1(config-VLAN 10)#ip address 192.168.1.126 255.255.255.128

SW1(config-VLAN 10)#interface vlan 20

SW1(config-VLAN 20)#ip address 192.168.1.254 255.255.255.128

SW1(config-VLAN 20)#interface vlan 30

SW1(config-VLAN 30)#ip address 192.168.2.11 255.255.255.0

SW1(config)#interface range f0/1-5

SW1(config-if-range)#switchport port-security

SW1(config-if-range)#switchport port-security maximum 4
SW1(config-if-range)#switchport port-security violation shutdown
SW1(config)#service dhcp
SW1(config)#ip dhcp pool vlan10
SW1(dhcp-config)#network 192.168.1.0 255.255.255.128
SW1(dhcp-config)#default-router 192.168.1.126
SW1(dhcp-config)#dns-server 8.8.8.8
SW1(dhcp-config)#exit
SW1(config)#ip dhcp pool vlan20
SW1(dhcp-config)#network 192.168.1.128 255.255.255.128
SW1(dhcp-config)#default-router 192.168.1.254
SW1(dhcp-config)#dns-server 8.8.8.8
SW1(dhcp-config)#exit
SW1(config)#router ospf 1
SW1(config-router)#router-id 5.5.5.5
SW1(config-router)#network 192.168.2.8 0.0.0.7   area 0
SW1(config-router)#network 192.168.1.0 0.0.0.127   area 0
SW1(config-router)#network 192.168.1.128 0.0.0.127 area 0
SW1(config)#ip route0.0.0.0 0.0.0.0   192.168.2.11

[SW2]

Ruijie>enable
Ruijie#configure   terminal
Enter configuration commands, one per line.   End with CNTL/Z.
Ruijie(config)#hostname SW2
SW2(config)#vlan range 10,20,30,40
SW2(config-vlan-range)#exit
SW2(config)#interface range f0/1-5
SW2(config-if-range)#switchport access vlan 10
SW2(config)#interface range f0/6-10
SW2(config-if-range)#switchport access vlan 20
SW2(config)#interface range f0/11-12
SW2(config-if-range)#switchport access vlan 40
SW2(config-if-range)#exit
SW2(config)#interface range f0/13-16

SW2(config-if-range)#port-group 1
SW2(config-if-range)#interface aggregate 1
SW2(config-AggregatePort 1)#switchport mode trunk
SW2(config)#interface vlan 10
SW2(config-VLAN 10)#ip address 192.168.1.125 255.255.255.128
SW2(config-VLAN 10)#interface vlan 20
SW2(config-VLAN 20)#ip address 192.168.1.253 255.255.255.128
SW2(config-VLAN 20)#interface vlan 30
SW2(config-VLAN 30)#ip address 192.168.2.11 255.255.255.248
SW2(config)#interface range f0/1-5
SW2(config-if-range)#switchport port-security
SW2(config-if-range)#switchport port-security maximum 4
SW2(config-if-range)#switchport port-security violation shutdown
SW2(config)#service dhcp
SW2(config)#ip dhcp pool vlan10
SW2(dhcp-config)#network 192.168.1.0 255.255.255.128
SW2(dhcp-config)#default-router 192.168.1.125
SW2(dhcp-config)#dns-server 8.8.8.8
SW2(config)#ip dhcp pool vlan20
SW2(dhcp-config)#network 192.168.1.128 255.255.255.128
SW2(dhcp-config)#default-router 192.168.1.253
SW2(dhcp-config)#dns-server 8.8.8.8
SW2(config)#router ospf 1
SW2(config-router)#router-id 6.6.6.6
SW2(config-router)#network 192.168.2.16 0.0.0.7 area 0
SW2(config-router)#network 192.168.1.0 0.0.0.127 area 0
SW2(config-router)#network 192.168.1.128 0.0.0.127 area 0
SW1(config)#ip route0.0.0.0 0.0.0.0   192.168.2.20

[R1]

R1>enable
R1#configure   terminal
Enter configuration commands, one per line.   End with CNTL/Z.
R1(config)#hostname R1
R1(config)#interface f0/0

R1(config-if-FastEthernet 0/0)#ip address 192.168.2.1 255.255.255.252

R1(config-if-FastEthernet 0/0)#exit

R1(config)#interface f0/1

R1(config-if-FastEthernet 0/1)#ip address 192.168.2.9 255.255.255.248

R1(config-if-FastEthernet 0/1)#ip nat inside

R1(config-if-FastEthernet 0/1)#vrrp 10 ip 192.168.2.11

R1(config-if-FastEthernet 0/1)#vrrp 10 priority 120

R1(config-if-FastEthernet 0/1)#interface f0/2

R1(config-if-FastEthernet 0/2)#ip address 192.168.2.17 255.255.255.248

R1(config-if-FastEthernet 0/2)#ip nat inside

R1(config-if-FastEthernet 0/2)#vrrp 20 ip 192.168.2.20

R1(config-if-FastEthernet 0/2)#interface s2/0

R1(config-if-Serial 2/0)#ip address 200.1.1.1 255.255.255.252

R1(config-if-Serial 2/0)#ip access-group 111 in

R1(config-if-Serial 2/0)#ip nat outside

R1(config-if-Serial 2/0)#exit

R1(config)#access-list 110 permit tcp any any eq 443

R1(config)#access-list 110 permit tcp any any eq 80

R1(config)#access-list 110 permit tcp any any eq 53

R1(config)#access-list 110 permit udp any any eq 53

R1(config)#access-list 111 deny tcp any any eq 138

R1(config)#access-list 111 deny tcp any any eq 139

R1(config)#ip route0.0.0.0 0.0.0.0   s 2/0

R1(config)#router ospf 1

R1(config-router)#router-id 1.1.1.1

R1(config-router)#network 192.168.2.0 0.0.0.255   area 0

R1(config-router)#network 172.16.2.8 0.0.0.3   area 0

R1(config-router)#network 172.16.2.16 0.0.0.3   area 0

R1(config-router)#default-information originate always

R1(config)#access-list 10 permit 192.168.1.0 0.0.0.255

R1(config)#ip nat inside source list 10 interface serial 2/0 overload

[R2]

Ruijie>enable

Ruijie#configure   terminal

Ruijie(config)#hostname R2
R2(config)#interface f0/0
R2(config-if-FastEthernet 0/0)#ip address 192.168.2.2 255.255.255.252
R2(config-if-FastEthernet 0/0)#ip nat inside
R2(config-if-FastEthernet 0/0)#interface   f0/1
R2(config-if-FastEthernet 0/1)#vrrp 10 ip 192.168.2.11
R2(config-if-FastEthernet 0/1)#ip address 192.168.2.10 255.255.255.248
R2(config-if-FastEthernet 0/1)#interface   f0/2
R2(config-if-FastEthernet 0/1)#ip   nat inside
R2(config-if-FastEthernet 0/2)#ip address 192.168.2.18 255.255.255.248
R2(config-if-FastEthernet 0/2)#vrrp 20 ip 192.168.2.20
R2(config-if-FastEthernet 0/1)#vrrp 20 priority 120
R2(config-if-FastEthernet 0/2)#interface   s2/0
R2(config-if-Serial 2/0)#ip   address   200.1.1.5   255.255.255.252
R2(config-if-Serial 2/0)#ip   nat outside
R2(config)#ip   route 0.0.0.0   0.0.0.0   s2/0
R2(config)#access-list 110 permit tcp any any eq 443
R2(config)#access-list 110 permit tcp any any eq 80
R2(config)#access-list 110 permit tcp any any eq 53
R2(config)#access-list 110 permit udp any any eq 53
R2(config)#access-list 111 deny tcp any any eq 138
R2(config)#access-list 111 deny tcp any any eq 139
R2(config)#access-list 121 permit ip 192.168.1.0 0.0.0.255 192.168.3.4 0.0.0.3
R2(config)#router ospf 1
R2(config-router)#router-id 2.2.2.2
R2(config-router)#network 192.168.2.0 0.0.0.3 area 0
R2(config-router)#network 192.168.2.8 0.0.0.7 area 0
R2(config-router)#network 192.168.2.16 0.0.0.7 area 0
R2(config-router)#default-information originate always
R2(config)#access-list 10 permit 192.168.1.0 0.0.0.255
R2(config)#ip nat inside source list 10 interface serial 2/0 overload
R2(config)#crypto isakmp policy 1
R2(isakmp-policy)#hash md5
R2(isakmp-policy)#authentication pre-share
R2(isakmp-policy)#encryption 3des
R2(config)#crypto isakmp key 0 ruijie address 200.1.1.9

R2(config)#crypto ipsec transform-set myset esp-des esp-md5-hmac

R2(config)#crypto map   mymap 1 ipsec-isakmp

R2(config-crypto-map)#set peer 200.1.1.9

R2(config-crypto-map)#match address 121

R2(config-crypto-map)#set transform-set myset

[R3]

Ruijie>enable

Ruijie#configure terminal

Enter configuration commands, one per line.   End with CNTL/Z.

Ruijie(config)#hostname R3

R3(config)#interface   f0/0

R3(config-if-FastEthernet 0/0)#ip address 200.1.1.9 255.255.255.252

R3(config-if-FastEthernet 0/0)#interface   f0/1

R3(config-if-FastEthernet 0/0)#ip nat outside

R3(config-if-FastEthernet 0/1)#ip address 192.168.3.1   255.255.255.252

R3(config-if-FastEthernet 0/1)#interface   f0/2

R3(config-if-FastEthernet 0/1)#ip   nat inside

R3(config-if-FastEthernet 0/2)#ip address 192.168.3.5   255.255.255.252

R2(config)#access-list 121 permit ip 192.168.3.4 0.0.0.3 192.168.1.0 0.0.0.255

R3(config)#ip nat inside source static tcp 192.168.3.6 80 interface f 0/0 80

R3(config)#ip nat inside source static tcp 192.168.3.6 21 interface f 0/0 21

R3(config)#crypto isakmp policy 1

R3(isakmp-policy)#hash md5

R3(isakmp-policy)#authentication pre-share

R3(isakmp-policy)#encryption 3des

R3(config)#crypto isakmp key 0 ruijie address 200.1.1.5

R3(config)#crypto ipsec transform-set myset esp-des esp-md5-hmac

R3(config)#crypto map   mymap 1 ipsec-isakmp

R3(config-crypto-map)#set peer 200.1.1.5

R3(config-crypto-map)#match address 121

R3(config-crypto-map)#set transform-set myset

R3(config)#router ospf 1

R3(config-router)#router-id 3.3.3.3

R3(config-router)#network 200.1.1.8   0.0.0.3   area 0

R3(config-router)#network 192.168.3.0 0.0.0.3   area 0

R3(config-router)#network 192.168.3.4 0.0.0.3   area 0

[R4]

Ruijie>enable

Ruijie#configure   terminal

Enter configuration commands, one per line.   End with CNTL/Z.

Ruijie(config)#hostname R4

R4(config)#interface s2/0

R4(config-if-Serial 2/0)#ip address 200.1.1.2   255.255.255.252

R4(config-if-Serial 3/0)#ip address 200.1.1.6   255.255.255.252

R4(config-if-Serial 3/0)#interface   f0/0

R4(config-if-FastEthernet 0/0)#ip address   200.1.1.10   255.255.255.252

R4(config-if-FastEthernet 0/0)#interface   f0/1

R4(config-if-FastEthernet 0/1)#ip address   200.1.1.14   255.255.255.0

[AP]

AP(config)#interface GigabitEthernet 0/1

AP(config-if- gigabitEthernet 0/1)#ip address 192.168.3.2 255.255.255.252

AP(config-if- gigabitEthernet 0/1)#speed 100

AP(config-if- gigabitEthernet 0/1)#exit

AP(config)#service dhcp

AP(config)#ip dhcp exclusive 192.168.3.9

AP(config)#ip dhcp pool pool1

AP(dhcp-config)#network 192.168.3.8   255.255.255.248

AP(dhcp-config)#default-router 192.168.3.9

AP(config)#dot11 wlan 10

AP(dot11-wlan-config)#vlan 1

AP(dot11-wlan-config)#no broadcast-ssid

AP(dot11-wlan-config)#ssid Ruijie

AP(dot11-wlan-config)#exit

AP(config)#interface Dot11radio 1/0

AP(config-if-Dot11radio 1/0)#encapsulation dot1Q 1

AP(config-if-Dot11radio 1/0)#mac-mode fat

AP(config-if-Dot11radio 1/0)#radio-type 802.11b

AP(config-if-Dot11radio 1/0)#channel 1

AP(config-if-Dot11radio 1/0)#wlan-id 10

AP(config-if-Dot11radio 2/0)#encapsulation dot1Q 1

AP(config-if-Dot11radio 2/0)#mac-mode fat

AP(config-if-Dot11radio 2/0)#radio-type 802.11a

AP(config-if-Dot11radio 2/0)#channel 149

AP(config-if-Dot11radio 2/0)#wlan-id 10

AP(config-if-Dot11radio 2/0)#exit

AP(config)#interface BVI 1

AP(config-if-BVI 1)#ip address 192.168.3.9   255.255.255.248

AP(config)#router ospf 10

AP(config-router)#router-id 7.7.7.7

AP(config-router)#network   192.168.3.0 0.0.0.3   area   0

AP(config-router)#network   192.168.3.8 0.0.0.7   area   0

AP(config-router)#exit

AP(config)#wlansec 10

AP(wlansec)#security static-wep-key encryption 40 ascii 1 12345

2012年青岛市中职技能大赛企业网搭建及应用模拟试题

一、竞赛环境

硬件环境：

设备类型	设备型号	设备数量（台）
路由器	RG-RSR20-04	4
三层交换机	RG-S3760E-24	2
防火墙	RG-WALL 160S	1
无线控制器	RG-WS3302	1
无线AP	RG-AP220E	1
计算机	—	4

软件环境：

软件名称	介质形式	软件数量
Windows 2003 Server 企业版	光盘	1
Windows XP Professional	光盘	1

二、网络拓扑

某市教育局信息中心使用的是锐捷网络的设备，下属的三个学校分别通过双线路连接到市局的核心交换机和备份的路由器，市局信息中心通过防火墙连接到 Internet 网络，负责全网的用户上网。其网络拓扑结构图如图 7-3-1 所示。

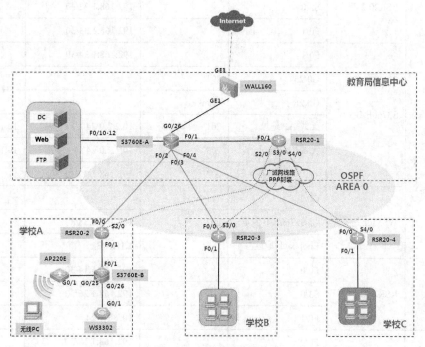

图 7-3-1　网络拓扑结构图

三、IP 规划（如表 7-2 所示）

表 7-3

网络区域	设备名称	设备接口	VLAN 号	IP 地址(段)	网关地址
教育局信息中心	WALL160S	GE0/3		200.1.1.1/24	200.1.1.2/24
		GE0/1		192.168.1.1/30	
	S3760E-A	G0/26		192.168.1.2/30	
		F0/1		192.168.1.5/30	
		F0/2		192.168.1.9/30	
		F0/3		192.168.1.13/30	
		F0/4		192.168.1.17/30	
		F0/10-12	101	10.1.1.0/24	10.1.1.1/24
	RSR20-1	F0/1		192.168.1.6/30	
		S2/0		192.168.1.21/30	
		S3/0		192.168.1.25/30	

续表

网络区域	设备名称	设备接口	VLAN 号	IP 地址(段)	网关地址
学校 A		S4/0		192.168.1.29/30	
学校 A	RSR20-2	F0/0		192.168.1.10/30	
学校 A	RSR20-2	S2/0		192.168.1.22/30	
学校 A	RSR20-2	F0/1		192.168.1.33/30	
学校 A	S3760E-B	F0/1		192.168.1.34/30	
学校 A	S3760E-B	G0/25	201		
学校 A	S3760E-B	G0/26	Trunk		
学校 A	S3760E-B	无线 AP 地址	201	10.2.2.0/24	10.2.2.1/24
学校 A	S3760E-B	无线用户地址	203	10.1.2.0/24	10.1.2.1/24
学校 A	S3760E-B	与无线控制器互联	202	192.168.1.37/30	
学校 A	WS3302	G0/1	Trunk		
学校 A	WS3302	与 S3760E-B 互联	202	192.168.1.38/30	
学校 A	WS3302	Loopback 地址		1.1.1.1/32	
学校 B	RSR20-3	F0/0		192.168.1.14/30	
学校 B	RSR20-3	S3/0		192.168.1.26/30	
学校 B	RSR20-3	F0/1		10.1.3.1/24	
学校 B	RSR20-4	F0/0		192.168.1.18/30	
学校 B	RSR20-4	S4/0		192.168.1.30/30	
学校 B	RSR20-4	F0/1		10.1.4.1/24	

四、试题内容

用户需求如下。

（1）网络底层配置。

根据图 7-3-1 所示，对网络设备的各接口、VLAN 等相关信息进行配置，使其能够正常通信。在学校路由器与 RSR20-1 之间以 PPP 协议进行封装。

（2）路由协议配置。

根据图 7-3-1 所示，配置全网的静态路由或者动态路由协议 OSPF，使网络互联互通。

要求实现学校 A、B、C 的主链路为各学校路由器到 S3760E-A 的线路，备份线路为各学校路由器到 RSR20-1 的线路。

要求将网络设备的各网段分别发布到 OSPF 中，要求能够互相访问。

（3）网络地址分配配置。

S3760E-A 所连接的服务器群地址为 10.1.1.0/24，属于 VLAN101，网关在 S3760-A 上

学校 B 用户分配地址为 10.1.3.0/24，网关为 10.1.3.1/24。

学校 C 用户分配地址为 10.1.4.0/24，网关为 10.1.4.1/24。

（4）链路安全配置。

要求在 S3760E-A 上 F0/20-24 设置端口安全，违背端口安全，采取行为 restrict，并设定安全地址个数为 5。

（5）出口防火墙配置。

配置安全策略最大限度地保证了内网和服务器群的安全。

创建时间访问控制列表，只有工作日（周一～周五）的工作时间（9:00～18:00）才能访问互联网。

在出口防火墙上做 NAT，将内网的 10.1.1.0/24、10.1.2.0/24、10.1.3.0/24、10.1.4.0/24 转换为外网的地址，地址池为 200.1.1.5～200.1.1.10。

（6）无线配置。

学校 A 的无线用户地址为 10.1.2.0/24 和无线 AP 的地址为 10.2.2.0/24，需要通过 DHCP SERVER 获取，服务在 S3760E-B 上启用，在无线 AP 的 DHCP 服务中需要将无线控制器的 loopback 地址 1.1.1.1 发送给无线 AP，使用 option138 功能。

在无线控制 WS3302 上需要对无线 AP 进行管理配置，建立 SSID 为 RUIJIE 的无线网络。允许广播 SSID，并配置 WEP 加密，设置其口令为 12345。

要求带有无线网卡的计算机能够在网卡中发现 SSID 为 ruijie 的无线服务，并且需要进行 WEP 认证后才能进入网络。

五．参考答案

[RSR20-1]

Ruijie>
Ruijie#configure terminal
Enter configuration commands, one per line.   End with CNTL/Z.
Ruijie(config)#hostname RSR20-1
RSR20-1(config)#interface f0/1
RSR20-1(config-if-FastEthernet 0/1)#ip address 192.168.1.6   255.255.255.252
RSR20-1(config-if-FastEthernet 0/1)#interface s2/0
RSR20-1(config-if-Serial 2/0)#ip address 192.168.1.21 255.255.255.252
RSR20-1(config-if-Serial 2/0)#interface s3/0
RSR20-1(config-if-Serial 3/0)#ip address 192.168.1.25 255.255.255.252
RSR20-1(config-if-Serial 3/0)#interface s4/0
RSR20-1(config-if-Serial 4/0)#ip address 192.168.1.29 255.255.255.252

RSR20-1(config)#router ospf 1

RSR20-1(config-router)#router-id 1.1.1.1

RSR20-1(config-router)#network 192.168.1.4 0.0.0.3 area 0

RSR20-1(config-router)#network 192.168.1.20 0.0.0.3 area 0

RSR20-1(config-router)#network 192.168.1.24 0.0.0.3 area 0

RSR20-1(config-router)#network 192.168.1.28 0.0.0.3 area 0

RSR20-1(config)#interface s 2/0

RSR20-1(config-if-Serial 2/0)#encapsulation ppp

RSR20-1(config-if-Serial 2/0)#ppp authentication pap

RSR20-1(config)#username ruijie password ruijie

[RSR20-2]

Ruijie>enable

Ruijie# configure terminal

Enter configuration commands, one per line.   End with CNTL/Z.

Ruijie(config)#hostname RSR20-2

RSR20-2(config)#interface f0/0

RSR20-2(config-if-FastEthernet 0/0)#ip address 192.168.1.10 255.255.255.252

RSR20-2(config-if-FastEthernet 0/0)#interface f0/1

RSR20-2(config-if-FastEthernet 0/1)#ip address 192.168.1.33 255.255.255.252

RSR20-2(config-if-FastEthernet 0/1)#interface s2/0

RSR20-2(config-if-Serial 2/0)#ip address 192.168.1.22 255.255.255.252

RSR20-2(config-if-Serial 2/0)#encapsulation ppp

RSR20-2(config-if-Serial 2/0)#ppp authentication pap

RSR20-2(config-if-Serial 2/0)#ppp pap sent-username ruijie password ruijie

[RSR20-3]

Ruijie>enable

Ruijie# configure terminal

Enter configuration commands, one per line.   End with CNTL/Z.

Ruijie(config)#hostname RSR20-3

RSR20-3(config)#interface f0/0

RSR20-3(config-if-FastEthernet 0/0)#ip address 192.168.1.14 255.255.255.252

RSR20-3(config-if-FastEthernet 0/0)#interface f0/1
RSR20-3(config-if-FastEthernet 0/1)#ip address 10.1.3.1 255.255.255.0
RSR20-3(config-if-FastEthernet 0/1)#interface s3/0
RSR20-3(config-if-Serial 3/0)#ip address 192.168.1.26 255.255.255.252
RSR20-3(config-if-Serial 3/0)#encapsulation ppp
RSR20-3(config-if-Serial 3/0)#ppp authentication pap
RSR20-3(config-if-Serial 3/0)#ppp pap sent-username ruijie password ruijie
RSR20-3(config-if-Serial 3/0)#exit
RSR20-3(config)#service dhcp
RSR20-3(config)#ip dhcp pool school_B
RSR20-3(dhcp-config)#network 10.1.3.0 255.255.255.0
RSR20-3(dhcp-config)#default-router 10.1.3.1

[RSR20-4]

Ruijie>enable
Ruijie#configure terminal
Enter configuration commands, one per line.   End with CNTL/Z.
Ruijie(config)#hostname RSR20-4
RSR20-4(config)#interface f0/0
RSR20-4(config-if-FastEthernet 0/0)#ip address 192.168.1.18 255.255.255.252
RSR20-4(config-if-FastEthernet 0/0)#interface f0/1
RSR20-4(config-if-FastEthernet 0/1)#ip address 10.1.4.1 255.255.255.0
RSR20-4(config-if-FastEthernet 0/1)#interface s4/0
RSR20-4(config-if-Serial 4/0)#ip address 192.168.1.30 255.255.255.252
RSR20-4(config-if-Serial 4/0)#encapsulation ppp
RSR20-4(config-if-Serial 4/0)#ppp authentication pap
RSR20-4(config-if-Serial 4/0)#ppp pap sent-username ruijie password ruijie
RSR20-4(config)#service dhcp
RSR20-4(config)#ip dhcp pool school_C
RSR20-4(dhcp-config)#network 10.1.4.0 255.255.255.0
RSR20-4(dhcp-config)#default-router 10.1.4.1

[S3760E-A]

Ruijie>enable

Ruijie#configure terminal

Enter configuration commands, one per line.   End with CNTL/Z.

Ruijie(config)#hostname S3760E-A

S3760E-A(config)#interface g0/26

S3760E-A(config-GigabitEthernet 0/26)#no switchport

S3760E-A(config-GigabitEthernet 0/26)#ip address 192.168.1.2 255.255.255.252

S3760E-A(config-GigabitEthernet 0/26)#interface f0/1

S3760E-A(config-FastEthernet 0/1)#no switchport

S3760E-A(config-FastEthernet 0/1)#ip address 192.168.1.5 255.255.255.0

S3760E-A(config-FastEthernet 0/1)#ip address 192.168.1.5 255.255.255.252

S3760E-A(config-FastEthernet 0/1)#interface f0/2

S3760E-A(config-FastEthernet 0/2)#no switchport S3760E-A(config-FastEthernet 0/2)#ip address 192.168.1.9 255.255.255.252

S3760E-A(config-FastEthernet 0/2)#interface f0/3

S3760E-A(config-FastEthernet 0/2)#no switchport

S3760E-A(config-FastEthernet 0/3)#ip address 192.168.1.13 255.255.255.252

S3760E-A(config-FastEthernet 0/3)#interface f0/4

S3760E-A(config-FastEthernet 0/4)#no switchport

S3760E-A(config-FastEthernet 0/4)#ip address 192.168.1.17 255.255.255.252

S3760E-A(config-FastEthernet 0/4)#exit

S3760E-A(config)#vlan 101

S3760E-A(config-vlan)#exit

S3760E-A(config)#interface vlan 101

S3760E-A(config-VLAN 101)#ip address 10.1.1.1 255.255.255.0

S3760E-A(config-VLAN 101)#exit

S3760E-A(config)#interface range f0/10-12

S3760E-A(config-if-range)#switchport access vlan 101

S3760E-A(config)#router ospf 1

S3760E-A(config-router)#router-id 5.5.5.5

S3760E-A(config-router)#network 192.168.1.0 0.0.0.3 area 0

S3760E-A(config-router)#network 192.168.1.4 0.0.0.3 area 0

S3760E-A(config-router)#network 192.168.1.8 0.0.0.3 area 0

S3760E-A(config-router)#network 192.168.1.12 0.0.0.3 area 0

S3760E-A(config-router)#network 192.168.1.17 0.0.0.3 area 0

S3760E-A(config-router)#network 10.1.1.0 0.0.0.255 area 0

S3760E-A(config)#interface range f0/20-24

S3760E-A(config-if-range)#switchport port-security
S3760E-A(config-if-range)#switchport port-security maximum 5
S3760E-A(config-if-range)#switchport port-security violation restrict

[S3760E-B]

Ruijie>
Ruijie>enable
Ruijie#configrue terminal
Enter configuration commands, one per line.　End with CNTL/Z.
Ruijie(config)#hostname S3760-B
S3760-B(config)#interface f0/1
S3760-B(config-FastEthernet 0/1)#no switchport
S3760-B(config-FastEthernet 0/1)#ip address 192.168.1.34 255.255.255.252
S3760-B(config-FastEthernet 0/1)#interface g0/26
S3760-B(config-GigabitEthernet 0/26)#switchport mode　trunk
S3760-B(config-GigabitEthernet 0/26)#exit
S3760-B(config)#vlan range 201,202,203
S3760-B(config-vlan-range)#exit
S3760-B(config)#interface g0/25
S3760-B(config-GigabitEthernet 0/25)#switchport access vlan 201
S3760-B(config-GigabitEthernet 0/25)#interface vlan 201
S3760-B(config-VLAN 201)#ip address 10.2.2.1 255.255.255.0
S3760-B(config-VLAN 201)#interface vlan 202
S3760-B(config-VLAN 202)#ip address 192.168.1.37 255.255.255.252
S3760-B(config)#interface vlan 203
S3760-B(config-VLAN 203)#ip address 10.1.2.1 255.255.255.0
S3760-B(config)#router ospf 1
S3760-B(config-router)#router-id 6.6.6.6
S3760-B(config-router)#network 192.168.1.32 0.0.0.3 area 0
S3760-B(config-router)#network 10.1.2.0 0.0.0.255 area 0
S3760-B(config-router)#network 10.2.2.0 0.0.0.255 area 0
S3760-B(config-router)#network 192.168.1.36 0.0.0.255 area 0
S3760-B(config)#service dhcp
S3760-B(config)#ip dhcp pool AP
S3760-B(dhcp-config)#network 10.2.2.0 255.255.255.0

S3760-B(dhcp-config)#default-router 10.2.2.1

S3760-B(dhcp-config)#option 138 ip 1.1.1.1

S3760-B(dhcp-config)#exit

S3760-B(config)#ip dhcp pool user

S3760-B(dhcp-config)#network 10.1.2.0 255.255.255.0

S3760-B(dhcp-config)#default-router 10.1.2.1

[WS3302]

Ruijie>

Ruijie>enable

Ruijie#configure terminal

Enter configuration commands, one per line.   End with CNTL/Z.

Ruijie(config)#hostname WS3302

WS3302(config)#interface loopback 0

WS3302(config-if-Loopback 0)#ip address 1.1.1.1 255.255.255.0

WS3302(config)#interface g0/1

WS3302(config-if-GigabitEthernet 0/1)#switchport   mode trunk

WS3302(config-if-GigabitEthernet 0/1)#vlan 202

WS3302(config-if-GigabitEthernet 0/1)#vlan 203

WS3302(config)#interface vlan 202

WS3302(config-if-VLAN 202)#ip address 192.168.1.38 255.255.255.252

WS3302(config-if-VLAN 203)#ip address 10.1.2.2 255.255.255.0

WS3302(config)#router ospf 1

WS3302(config-router)#router-id 7.7.7.7

WS3302(config-router)#network 1.1.1.1 0.0.0.0 area 0

WS3302(config-router)#network 192.168.1.36 0.0.0.3 area 0

WS3302(config)#wlan-config 1 RUIJIE

WS3302(wlansec)#security static-wep-key authentication share-key

WS3302(wlansec)#security static-wep-key encryption 40 ascii 1 12345

WS3302(config)#ap-group ap

WS3302(config-ap-group)#interface-mapping 1 203

WS3302(wlansec)#ap-config all

WS3302(config-ap)#ap-group ap

[WALL160S]

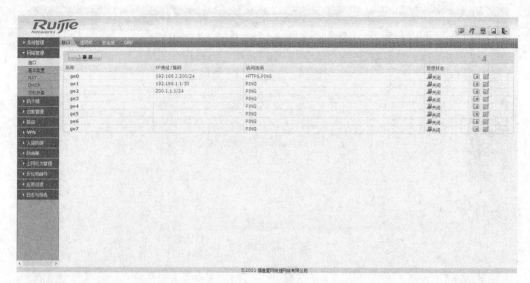

图 7-3-2 配置接口

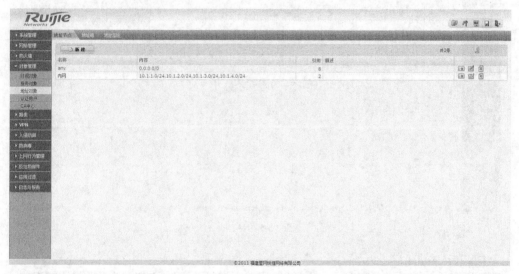

图 7-3-3 地址对象

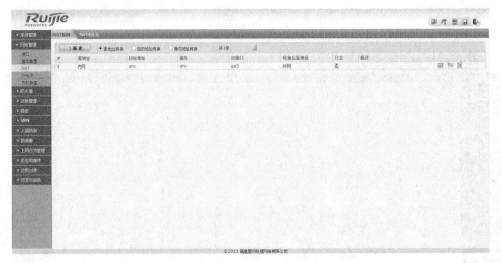

图 7-3-4　时间对象

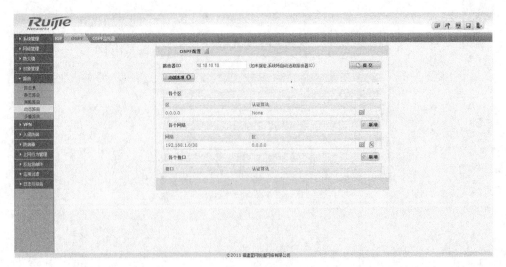

图 7-3-5　NAT

图 7-3-6　动态路由

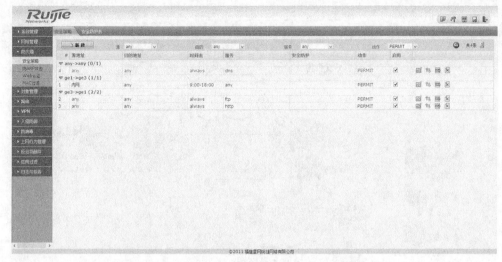

图 7-3-7 安全策略

2013年青岛市中职技能大赛企业网搭建及应用模拟试题

一、网络部分

硬件环境

设备类型	设备型号	设备数量（台）
路由器	RG-RSR20-18	4
三层交换机	RG-S3760E-24	2
二层交换机	RG-S2628G-I	2
防火墙	RG-WALL 160S	1
无线控制器	RG-WS3302	1
无线 AP	RG-AP220E	1
以太网供电	RG-E-130	1
计算机	—	4

二、网络拓扑

某市教育局信息中心使用的是锐捷网络的设备，下属的两个学校 A、B 分别通过专线连接到市局的核心路由器，同时某分支机构通过 Internet 线路与教育局网络中心通过 IPSEC 嵌套 GRE VPN 互联，与学校 A 通过 GRE VPN 互联，学校 B 与学校 A 通过专线互联以访问学校 A 的开放资源，市局信息中心通过路由器连接到 Internet 网络，负责教育局全网及学校 B 的用户上网。其网络拓扑结构图如图 7-4-1 所示：

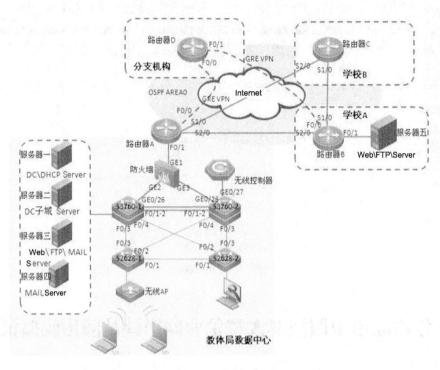

图 7-4-1　网络拓扑结构图

三、IP 规划（如表 7-4 所示）

网络区域	设备名称	设备接口	VLAN 号	IP 地址(段)	网关地址
教育局信息中心	WALL160S	GE0/3、GE0/2	VLAN200	192.168.200.1/24	
		GE0/1		192.168.1.1/30	
	S3760E-A	VLAN10		192.168.10.1/24	192.168.10.254
		VLAN20		192.168.20.1/24	192.168.20.254
		VLAN30		192.168.30.1/24	192.168.30.254
		VLAN40		192.168.40.1/24	192.168.40.254
		VLAN100		192.168.100.253/24	192.168.100.254/24
		VLAN200		192.168.200.2/24	
	路由器 A	F0/0		200.1.1.1/24	200.1.1.254/24
		F0/1		192.168.1.2/30	
		S1/0		172.16.1.1/30	
		S2/0		172.16.2.1/30	
		TUNNEL0		172.16.30.1/30	
		LOOPBACK0		8.8.8.1/24	
	S3760E-B	VLAN10		192.168.10.2/24	192.168.10.254
		VLAN20		192.168.20.2/24	192.168.20.254

续表

网络区域	设备名称	设备接口	VLAN 号	IP 地址(段)	网关地址
		VLAN30		192.168.30.2/24	192.168.30.254
		VLAN40		192.168.40.2/24	192.168.40.254
		VLAN100		192.168.100.252/24	192.168.100.254/24
		VLAN200		192.168.200.3/24	
		无线 AP 地址	201	10.2.2.0/24	10.2.2.1/24
		无线用户地址	203	10.1.2.0/24	10.1.2.254/24
		与无线控制器互联	202	192.168.1.37/30	
	WS3302	G0/1	Trunk		
		与 S3760E-B 互联	202	192.168.1.38/30	
		Loopback 地址		1.1.1.1/32	
学校 A	路由器 B	F0/0		201.1.1.2/24	
		F0/1		192.168.252.1/24	
		S1/0		172.16.3.1/30	
		S2/0		172.16.2.2/30	
		TUNNEL1		172.16.40.1/30	
		LOOPBACK0		9.9.9.1/24	
学校 B	路由器 C	S1/0		172.16.3.2/30	
		S2/0		172.16.1.2/30	
		LOOPBACK0		11.11.11.1/24	
分支机构	路由器 D	F0/0		200.1.1.2/24	
		F0/1		201.1.1.2/24	
		LOOPBACK0		12.12.12.1/24	
		TUNNEL0		172.16.30.2/30	
		TUNNEL1		172.16.40.2/30	
服务器	服务器一	DC/DNS/DHCP		192.168.100.1/24	
	服务器二	DC 子域		192.168.100.2/24	
	服务器三	Web/FTP(Linux)		192.168.100.3/24	
	服务器四	MIAL		192.168.100.4/24	
	服务器五	Web/FTP(Windows)		192.168.252.2/24	
	配置机地址			根据赛场环境现场确定	

四、试题内容

用户需求如下。

（1）网络底层配置。（6 分）

根据图 7-4-1 所示，对网络设备的各接口、VLAN 等相关信息进行配置，使其能够正常通信。在中心与学校路由器之间的专线广域网链路使用 PPP 协议进行封装。

（2）路由协议配置。（20 分）

根据图 7-4-1 所示，配置全网的静态路由或者动态路由协议 OSPF，使网络互联互通。

要求实现学校 B 访问教育局中心服务器的主链路为学校路由器到路由器 A 的线路，备份线路为学校 B 路由器到学校 A 的线路；同时实现学校 B 访问学校 A 服务器的主链路为学校路由器到路由器 B 的线路，备份线路为学校 B 路由器到教育局中心的线路。学校 B 通过局中心路由器访问 Internet，请设计 NAT 规划。转换地址为 200.1.1.12。

要求将网络设备的各网段分别发布到 OSPF 中，要求能够互相访问。

局中心 Web 服务器的地址为 192.168.100.3,将 Web 服务发布到外网上，外网 IP 为 200.1.1.11，不允许外网用户通过 Internet 访问 FTP 服务器。

S3760E-A 与 S3760E-B 之间运行 VRRP，S3760E-A 为 VLAN10、VLAN20 的 MASTER，S3760E-B 为 VLAN30、VLAN40 的 MASTER。

（3）网络地址分配配置。（8 分）

按表 7-4 要求进行正确的地址配置，并创建相应的 VLAN。

（4）链路安全配置。（6 分）

要求在两台 S2628 上的 F0/20-24 设置端口安全，违背端口安全，采取行为 restrict。指定最大允许学多少个地址数为 5。

（5）出口防火墙配置。（15 分）

配置安全策略最大限度地保证了内网和服务器群的安全。

在防火墙上做 NAT，将教育局内网的 VLAN10-40 及无线用户地址转换为外网的地址，地址池为 200.1.1.5～200.1.1.10。

创建时间访问控制列表，只有工作日（周一～周五）的工作时间（9:00～18:00）才能访问互联网。

（6）无线配置（15 分）

无线 AP 的地址为 10.2.2.0/24，需要通过 DHCP SERVER 获取，服务在 S3760A 上启用，在无线 AP 的 DHCP 服务中需要将无线控制器的 loopback 地址 1.1.1.1 发送给无线 AP，使用 option138 功能。无线用户地址为 10.1.2.0/24，需要通过 DHCP SERVER 获取，服务在 DHCP 服务器上启用，网关地址为 10.1.2.254，DNS 地址为内网 DNS 服务器地址，辅助 DNS 地址为 202.102.134.68。

在无线控制 WS3302 上需要对无线 AP 进行管理配置，建立 SSID 为 RUIJIE 并且允许广播，配置 WPA2 加密，设置其口令为 1234567890。

要求带有无线网卡的计算机能够在网卡中发现 SSID 为 ruijie 的无线服务，并且需要进行

WPA2 认证后才能进入网络。

五. 参考答案

[S3760E-A]

Ruijie>enable

Ruijie#configure terminal

Enter configuration commands, one per line.   End with CNTL/Z.

Ruijie(config)#hostname S3760E-A

S3760E-A(config)#vlan range 10,20,30,40,100,200

S3760E-A(config-vlan-range)#exit

S3760E-A(config)#interface vlan 10

S3760E-A(config-VLAN 10)#ip address 192.168.10.1 255.255.255.0

S3760E-A(config-VLAN 10)#vrrp 10 ip 192.168.10.254

S3760E-A(config-VLAN 10)#vrrp 10 priority 120

S3760E-A(config-VLAN 10)#interface vlan 20

S3760E-A(config-VLAN 20)#ip address 192.168.20.1 255.255.255.0

S3760E-A(config-VLAN 10)#vrrp 20 ip 192.168.20.254

S3760E-A(config-VLAN 10)#vrrp 20 priority 120

S3760E-A(config-VLAN 20)#interface vlan 30

S3760E-A(config-VLAN 30)#ip address 192.168.30.1 255.255.255.0

S3760E-A(config-VLAN 10)#vrrp 30 ip 192.168.30.254

S3760E-A(config-VLAN 30)#interface vlan 40

S3760E-A(config-VLAN 40)#ip address 192.168.40.1 255.255.255.0

S3760E-A(config-VLAN 10)#vrrp 40 ip 192.168.40.254

S3760E-A(config-VLAN 40)#interface vlan 100

S3760E-A(config-VLAN 100)#ip address 192.168.100.253 255.255.255.0

S3760E-A(config-VLAN 100)#interface vlan 200

S3760E-A(config-VLAN 200)#ip address 192.168.200.2 255.255.255.0

S3760E-A(config)#interface range f 0/1-2

S3760E-A(config-if-range)#port-group 1

S3760E-A(config-if-range)#interface aggregate 1

S3760E-A(config-AggregatePort 1)#switchport mode trunk

S3760E-A(config)#interface range f 0/3-4

S3760E-A(config-if-range)#switchport mode trunk

S3760E-A(config)#interface g0/26

S3760E-A(config-GigabitEthernet 0/26)#

S3760E-A(config-GigabitEthernet 0/26)#switchport access vlan 200

S3760E-A(config)#spanning-tree

S3760E-A(config)#spanning-tree mst configuration

S3760E-A(config-mst)#instance 10 vlan 10,20

S3760E-A(config-mst)#instance 20 vlan 30,40

S3760E-A(config)#spanning-tree mst 10 priority 4096

S3760E-A(config)#spanning-tree mst 20 priority 8192

S3760E-A(config)#router ospf 1

S3760E-A(config-router)#

S3760E-A(config-router)#network 192.168.10.00.0.0.255 area 0

S3760E-A(config-router)#network 192.168.20.00.0.0.255 area 0

S3760E-A(config-router)#network 192.168.30.00.0.0.255 area 0

S3760E-A(config-router)#network 192.168.40.00.0.0.255 area 0

S3760E-A(config-router)#network 192.168.100.00.0.0.255 area 0

S3760E-A(config-router)#network 192.168.200.00.0.0.255 area 0

S3760E-A(config)#service dhcp

S3760E-A(config)#ip dhcp pool ap

S3760E-A(dhcp-config)#network 10.2.2.0 255.255.255.0

S3760E-A(dhcp-config)#default-router 10.2.2.1

S3760E-A(dhcp-config)#option 138 ip 20-20-20-20

S3760E-A(config)#ip dhcp pool user

S3760E-A(dhcp-config)#network 10.1.2.0 255.255.255.0

S3760E-A(dhcp-config)#default-router 10.1.2.1

[S3760E-B]

Ruijie>enable

Ruijie#configure terminal

Enter configuration commands, one per line.   End with CNTL/Z.

Ruijie(config)#hostname S3760E-B

S3760E-B(config)#vlan range 10,20,30,40,100,200,201,202,203

S3760E-B(config-vlan-range)#exit

S3760E-B(config)#interface vlan 10

S3760E-B(config-VLAN 10)#ip address 192.168.10.2 255.255.255.0

```
S3760E-B(config-VLAN 10)#vrrp 10 ip 192.168.10.254
S3760E-B(config-VLAN 10)#interface vlan 20
S3760E-B(config-VLAN 20)#ip address 192.168.20.2 255.255.255.0
S3760E-B(config-VLAN 20)#vrrp 20 ip 192.168.20.254
S3760E-B(config-VLAN 20)#interface vlan 30
S3760E-B(config-VLAN 30)#ip address 192.168.30.2 255.255.255.0
S3760E-B(config-VLAN 40)#vrrp 30 ip 192.168.30.254
S3760E-B(config-VLAN 40)#vrrp 30 priority 120
S3760E-B(config-VLAN 30)#interface vlan 40
S3760E-B(config-VLAN 40)#ip address 192.168.40.2 255.255.255.0
S3760E-B(config-VLAN 40)#vrrp 40 ip 192.168.40.254
S3760E-B(config-VLAN 40)#vrrp 40 priority 120
S3760E-B(config-VLAN 40)#interface vlan 100
S3760E-B(config-VLAN 100)#ip address 192.168.100.252 255.255.255.0
S3760E-B(config-VLAN 100)#interface vlan 200
S3760E-B(config-VLAN 200)#ip address 192.168.200.3 255.255.255.0
S3760-B(config)#interface vlan 201
S3760-B(config-VLAN 201)#exit
S3760-B(config-VLAN 201)#ip address 10.2.2.1 255.255.255.0
S3760-B(config-VLAN 201)#interface vlan 203
S3760-B(config-VLAN 203)#ip address 10.1.2.1 255.255.255.0
S3760-B(config-VLAN 203)#interface vlan 202
S3760-B(config-VLAN 202)#ip address 192.168.1.37 255.255.255.252
S3760-B(config-VLAN 202)#exit
S3760E-B(config)#interface range f 0/1-2
S3760E-B(config-if-range)#port-group 1
S3760E-B(config-if-range)#interface aggregate 1
S3760E-B(config-AggregatePort 1)#switchport mode trunk
S3760E-B(config)#interface range f 0/3-4
S3760E-B(config-if-range)#switchport mode trunk
S3760E-B(config)#interface g0/26
S3760E-B(config-GigabitEthernet 0/26)#switchport access vlan 200
S3760E-B(config)#spanning-tree
S3760E-B(config)#spanning-tree mst configuration
S3760E-B(config-mst)#instance 10 vlan 10,20
S3760E-B(config-mst)#instance 20 vlan 30,40
```

S3760E-B(config)#spanning-tree mst 10 priority 8192

S3760E-B(config)#spanning-tree mst 20 priority 4096

S3760E-B(config)#router ospf 1

S3760E-B(config-router)#

S3760E-B(config-router)#network 192.168.10.0 0.0.0.255 area 0

S3760E-B(config-router)#network 192.168.20.0 0.0.0.255 area 0

S3760E-B(config-router)#network 192.168.30.0 0.0.0.255 area 0

S3760E-B(config-router)#network 192.168.40.0 0.0.0.255 area 0

S3760E-B(config-router)#network 192.168.100.0 0.0.0.255 area 0

S3760E-B(config-router)#network 192.168.200.0 0.0.0.255 area 0

S3760-B(config-router)#network 10.2.2.0 0.0.0.255 area 0

S3760-B(config-router)#network 10.1.2.0 0.0.0.255 area 0

S3760-B(config-router)#network 192.168.1.36 0.0.0.3 area 0

[RSR20-A]

Ruijie>enable

Ruijie#configure terminal

Enter configuration commands, one per line.   End with CNTL/Z.

Ruijie(config)#hostname RSR20-A

RSR20-A(config)#interface f 0/0

RSR20-A(config-if-FastEthernet 0/0)#ip address 200.1.1.1 255.255.255.0

RSR20-A(config-if-FastEthernet 0/0)#interface f0/1

RSR20-A(config-if-FastEthernet 0/1)#ip address 192.168.1.2 255.255.255.252

RSR20-A(config-if-FastEthernet 0/1)#interface s2/0

RSR20-A(config-if-Serial 2/0)#ip address 172.16.2.1 255.255.255.252

RSR20-A(config-if-Serial 2/0)#interface tunnel 0

RSR20-A(config-if-Tunnel 0)#ip address 172.16.30.1 255.255.255.252

RSR20-A(config-if-Tunnel 0)#tunnel source f0/0

RSR20-A(config-if-Tunnel 0)#tunnel destination 200.1.1.2

RSR20-A(config-if-Tunnel 0)#exit

RSR20-A(config)#interface serial 1/0

RSR20-A(config-if-Serial 1/0)#ip address 172.16.1.1 255.255.255.252

RSR20-A(config-if-Serial 1/0)#interface loopback 0

RSR20-A(config-if-Loopback 0)#ip address 8-8-8-1 255.255.255.0

RSR20-A(config)#ip route 0.0.0.0 0.0.0.0 serial 1/0

RSR20-A(config)#router ospf 1
RSR20-A(config-router)#network 192.168.1.00.0.0.3 area 0
RSR20-A(config-router)#network 172.16.2.00.0.0.3 area 0
RSR20-A(config-router)#default-information originate always
RSR20-A(config)#crypto isakmp policy 1
RSR20-A(isakmp-policy)#hash md5
RSR20-A(isakmp-policy)#authentication pre-share
RSR20-A(isakmp-policy)#encryption 3des
RSR20-A(config)#crypto isakmp key 0 ruijie address 200.1.1.2
RSR20-A(config)#crypto ipsec transform-set myset esp-des esp-md5-hmac
RSR20-A(config)#crypto map   mymap 1 ipsec-isakmp
RSR20-A(config-crypto-map)#set peer 200.1.1.2
RSR20-A(config-crypto-map)#match address 110
RSR20-A(config-crypto-map)#set transform-set myset
RSR20-A(config)#access-list 110 permit ip host 200.1.1.1 host 200.1.1.2

[RSR20-B]

Ruijie>enable
Ruijie#configure terminal
Enter configuration commands, one per line.   End with CNTL/Z.
Ruijie(config)#hostname RSR20-B
RSR20-B(config)#interface f0/0
RSR20-B(config-if-FastEthernet 0/0)#ip address 201.1.1.2 255.255.255.0
RSR20-B(config-if-FastEthernet 0/1)#interface s1/0
RSR20-B(config-if-Serial 1/0)#ip address 172.16.3.1 255.255.255.252
RSR20-B(config)#interface s 2/0
RSR20-B(config-if-Serial 2/0)#ip address 172.16.2.2 255.255.255.252
RSR20-B(config)#interface tunnel 0
RSR20-B(config-if-Tunnel 0)#ip address 172.16.40.1 255.255.255.252
RSR20-B(config-if-Tunnel 0)#tunnel source f 0/0
RSR20-B(config-if-Tunnel 0)#tunnel destination 201.1.1.1
RSR20-B(config-if-Tunnel 0)#interface loopback 0
RSR20-B(config-if-Loopback 0)#ip address 9.9.9.1 255.255.255.0
RSR20-B(config)#router ospf 1
RSR20-B(config-router)#network 172.16.3.00.0.0.3 area 0

RSR20-B(config-router)#network 192.168.252.0 0.0.0.255 area 0
RSR20-B(config-router)#network 172.16.2.0 0.0.0.3 area 0
RSR20-B(config-router)#network 172.16.40.0 0.0.0.3 area 0
RSR20-B(config-router)#network 9.9.9.0 0.0.0.255 area 0

[RSR20-C]

Ruijie>enable
Ruijie#configure terminal
Enter configuration commands, one per line.   End with CNTL/Z.
Ruijie(config)#hostname RSR20-C
RSR20-C(config)#interface s1/0
RSR20-C(config-if-Serial 1/0)#ip address 172.16.3.2 255.255.255.252
RSR20-C(config)#interface s2/0
RSR20-C(config-if-Serial 2/0)#ip address 172.16.1.2 255.255.255.252
RSR20-C(config)#interface loopback 0
RSR20-C(config-if-Loopback 0)#ip address 11.11.11.1 255.255.255.0
RSR20-C(config)#router ospf 1
RSR20-C(config-router)#network 172.16.3.0 0.0.0.3 area 0
RSR20-C(config-router)#network 172.16.2.0 0.0.0.3 area 0
RSR20-C(config-router)#network 11.11.11.0 0.0.0.255 area 0

[RSR20-D]

Ruijie>enable
Ruijie#configure terminal
Enter configuration commands, one per line.   End with CNTL/Z.
Ruijie(config)#hostname RSR20-D
RSR20-D(config)#interface f0/0
RSR20-D(config-if-FastEthernet 0/0)#ip address 200.1.1.2 255.255.255.0
RSR20-D(config-if-FastEthernet 0/0)#interface f0/1
RSR20-D(config-if-FastEthernet 0/1)#ip address 201.1.1.2 255.255.255.0
RSR20-D(config-if-FastEthernet 0/1)#no shutdown
RSR20-D(config-if-FastEthernet 0/1)#exit

RSR20-D(config-if-FastEthernet 0/1)#interface loopback 0

RSR20-D(config-if-Loopback 0)#ip address 12.12.12.1 255.255.255.0

SR20-D(config-if-Loopback 0)#interface tunnel 0

RSR20-D(config-if-Tunnel 0)#ip address 172.16.30.2 255.255.255.0

RSR20-D(config-if-Tunnel 0)#tunnel source f 0/0

RSR20-D(config-if-Tunnel 0)#tunnel destination 200.1.1.1

RSR20-D(config-if-Tunnel 0)#exit

RSR20-D(config)#interface tunnel 1

RSR20-D(config-if-Tunnel 1)#ip address 172.16.40.2 255.255.255.252

RSR20-D(config-if-Tunnel 1)#tunnel source f0/1

RSR20-D(config-if-Tunnel 1)#tunnel destination 201.1.1.1

RSR20-D(config)#router ospf 1

RSR20-D(config-router)#network 172.16.30.00.0.0.3 area 0

RSR20-D(config-router)#network 172.16.40.00.0.0.3 area 0

RSR20-D(config-router)#network 12.12.12.10.0.0.255 area 0

[S2628-1]

Ruijie>

Ruijie>enable

Ruijie#configure terminal

Ruijie(config)#hostname S2126-1

S2126-1(config)#vlan ran 10,20,30,40,50,100,200

S2126-1(config-vlan-range)#exit

S2126-1(config)#interface range f0/1-3

S2126-1(config-if-range)#switchport mode trunk

S2126-1(config)#spanning-tree

S2126-1(config)#spanning-tree mst configuration

S2126-1(config-mst)#instance 10 vlan 10,20

S2126-1(config-mst)#instance 20 vlan 30,40

S2126-1(config)#interface range fastEthernet 0/20-24

S2126-1(config-if-range)#switchport port-security

S2126-1(config-if-range)#switchport port-security maximum 5

S2126-1(config-if-range)#switchport port-security violation restrict

[S2628-2]

Ruijie>
Ruijie>enable
Ruijie#configure terminal
Ruijie(config)#hostname S2126-2
S2126-2(config)#vlan range 10,20,30,40,50,100,200
S2126-2(config-vlan-range)#exit
S2126-2(config)#interface range f 0/1-3
S2126-2(config-if-range)#switchport mode trunk
S2126-2(config)#spanning-tree
S2126-2(config)#spanning-tree mst configuration
S2126-2(config-mst)#instance 10 vlan 10,20
S2126-2(config-mst)#instance 20 vlan 30,40
S2126-2(config)#interface range fastEthernet 0/20-24
S2126-2(config-if-range)#switchport port-security
S2126-2(config-if-range)#switchport port-security maximum 5
S2126-2(config-if-range)#switchport port-security violation restrict

[WS3302]

Ruijie>
Ruijie>enable
Ruijie#configure terminal
Enter configuration commands, one per line.   End with CNTL/Z.
Ruijie(config)#hostname WS3302
WS3302(config)#interface loopback 0
WS3302(config-if-Loopback 0)#ip address 20-20-20-20 255.255.255.0
WS3302(config)#interface g 0/1
WS3302(config-if-GigabitEthernet 0/1)#switchport   mode trunk
WS3302(config-if-GigabitEthernet 0/1)#vlan 202
WS3302(config-if-GigabitEthernet 0/1)#vlan 203
WS3302(config)#interface vlan 202
WS3302(config-if-VLAN 202)#ip address 192.168.1.38 255.255.255.252
WS3302(config-if-VLAN 203)#ip address 10.1.2.2 255.255.255.0

WS3302(config)#router ospf 1

WS3302(config-router)#network 1.1.1.10.0.0.0 area 0

WS3302(config-router)#network 192.168.1.360.0.0.3 area 0

WS3302(config)#wlan-config 1 RUIJIE

WS3302(wlansec)#security rsn enable

WS3302(wlansec)#security rsn akm psk enable

WS3302(wlansec)#security rsn akm psk set-key ascii 12345

WS3302(wlansec)#security rsn ciphers tkip enable

WS3302(config)#ap-group ap

WS3302(config-ap-group)#interface-mapping 1 203

WS3302(wlansec)#ap-config all

WS3302(config-ap)#ap-group ap

[WALL160S]

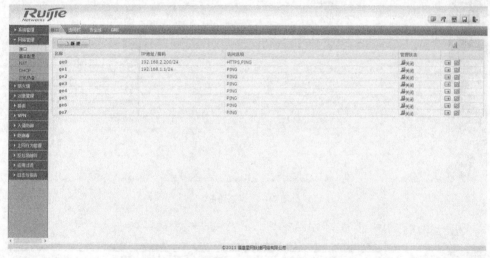

图 7-4-2　接口

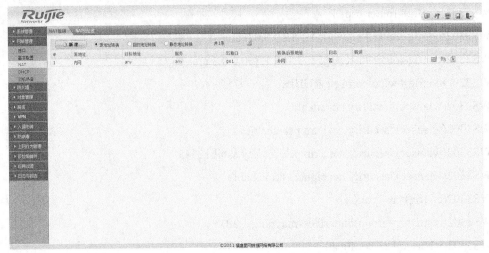

图 7-4-3　NAT

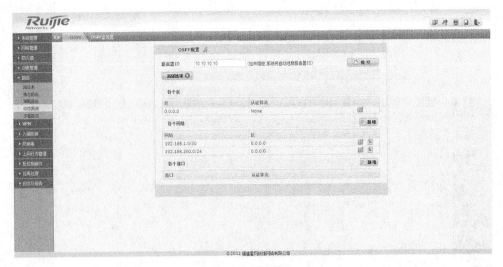

图 7-4-4　动态路由

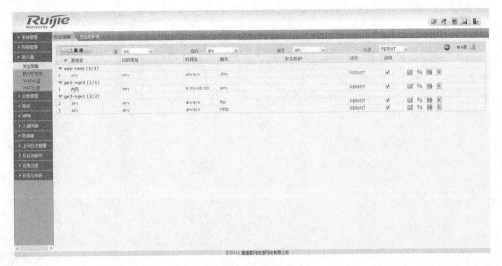

图 7-4-5　安全策略